Also by Klaus Truemper

Brain Science
Subconscious Blunders
Artificial Intelligence
Wittgenstein and Brain Science
Magic, Error, and Terror

Mathematics
Key Concepts of Mathematics
The Daring Invention of Logarithm Tables
(English and German edition)
The Construction of Mathematics

Technical
Logic-based Intelligent Systems
Effective Logic Computation
Matroid Theory

Edited by Ingrid and Klaus Truemper

F. Hülster *Introduction to Wittgenstein's Tractatus Logico-Philosophicus*
(English and German edition)

F. Hülster *Berlin 1945: Surviving the Collapse*

LESSONS FROM PILOTING FOR 45 YEARS

KLAUS TRUEMPER

Copyright © 2024 by Klaus Truemper

Softcover published by Leibniz Company
2304 Cliffside Drive
Plano, Texas, 75023
USA

Original edition 2024
Updated edition 2025

All rights reserved.

No part of this book may be reproduced, or stored in a retrieval system, or transmitted in any form or by any means, electronic, mechanical, photocopying, recording, or otherwise, without express permission of the publisher.

The book is typeset in LaTeX using the Tufte-style book class, which was inspired by the work of Edward R. Tufte and Richard Feynman.

Credits:
Almost all photos and maps are by the author. The Notes section lists the exceptions.

Book cover:
Photo by the author after an early-morning takeoff from Deming, New Mexico.

Library of Congress Cataloging-in-Publication Data
Truemper, Klaus, 1942–

Lessons from Piloting for 45 Years
ISBN 978-0-9991402-9-1
1. Piloting. 2. Safety.

There are old pilots and there are bold pilots, but there are no old, bold pilots.

Charles L. Wright
L & H Aircraft Corporation, 1931

Contents

1   *Introduction*   1
      The Trip   2
      The Decisions   2

2   *Planes and Friends*   5
      Friends   8

Part I   Weather   13

3   *Headwind*   15
      Another Surprise   19
      The Lesson   20

4   *Sully's Rule*   21
      Sully's Rule   23

5   *Virga Rain*   25
      The Lesson   26

6   *Mountain Air*   27
      The Lesson   28
      Additional Recommendation   29

7   *More on Headwind*   31
      The Solution   32

8 *Sudden Clouds*   35
    The Lesson   37

9 *Above the Clouds*   38
    VFR Cloud Conditions   39
    Cloud Types and Ceiling   39
    Consequences of the Cloud Rules   40
    VFR Over-the-Top   41
    Equipment   42
    Decision to Start VFR Over-the-Top   43
    Decision to Continue VFR Over-the-Top   43
    Special Aspects   44
    Conclusion   45

10 *Below the Clouds*   46
    Forecast Evaluation   47

11 *Thunderstorms*   49
    Types of Thunderstorms   50
    Planning for Thunderstorms   51

12 *Managing Thunderstorms*   52
    Conclusions   54

13 *Hot Air on the Ground*   56
    The Lesson   57
    Computing Density Altitude   58
    Simplified Rotax Procedure   59
    Engine Performance   59
    Takeoff rule for Zenith 601HDS N314LB   59

14 *Hot Air Up-High*   60
   The Trip   60
   Analysis   63
   The Lesson   64
   Additional Recommendation   64

Part II Fundamentals   65

15 *Traffic Pattern*   67
   The Lesson   68

16 *Landings*   70

17 *Distractions*   72
   The Lesson   73

18 *Dangerous Airports*   75
   Analysis   77
   Conclusion   77

Part III Emergencies   79

19 *Exhaust System*   81
   Puzzling Symptoms   81
   Taking Action   82
   Subsequent Contact with ATC   84
   The ATC Story   84
   Insight from the Zoom Session   86
   What's Wrong With the Engine?   87

    Analysis   88
    Repair   89
    What's to Be Learned?   90
    A Well-deserved Award   90

20 *Burning Smell*   92
    The Problem   92
    Conclusions   93

21 *Almost Out of Fuel*   95
    Choices   95
    Low-Fuel Management   96

22 *Excessive Oil Pressure*   98
    Analysis   98
    Conclusion   99

23 *Pilot Incapacitation*   101
    Checklist for Pilot Incapacitation   101
    Instructing the Copilot/Passenger   102

Part IV Good Practices   103

24 *Maintenance*   105

25 *Your Body and Soul*   107

Part V Joy of Flying   109

26 *Mountains*   111
    Caution   116

27 Rivers   117

Part VI   123

  Epilogue   125

  Appendix A: VFR Flight   128

  Glossary of Terms   130

  Notes   132

  Bibliography   136

  Acknowledgements   137

  Index   138

1

Introduction

May 8, 1988, 3 pm

A black wall of rain is moving toward the Weiss airport in Fenton, Missouri like a monstrous steamroller. It's about to flood the airport where I had landed a few minutes ago and was about to clean the bugs off the airplane. Instead, I rush into the terminal as a torrent of rain beats down.

The anemometer inside the building shows wind from the south at 25 kts. During the next 30 seconds, the wind direction swivels from south to north while the wind speed stays at 25 kts. If I had arrived just 10 minutes later, I would have flown into this tremendous downpour and been caught in the turbulent wind shift. No doubt, it would have been a deadly crash.

Yet, during the entire cross-country flight that day—starting from the Marion airport C17 in Eastern Iowa and ending at the Weiss airport[1] 3WE in Fenton southwest of St. Louis—all decisions seemed perfect. How is that colossal failure possible?

The Trip

The airplane of that fateful trip: a Grumman AA1-C, tail number N9782U. Taking off from the Marion airport midmorning, the flight proceeds southbound along the western shore of the Mississippi. There are some delays—to be discussed in a moment—but eventually I arrive at the Weiss airport. Buffeted by gusty winds coming from the south, I manage a reasonably smooth landing and tie down the airplane. Shortly after, the massive black wall of rain hits.

The Decisions

In 1988, FSS (Flight Service Station) briefers supplied weather information for flight planning on the telephone. En route, FSS flight specialists provided limited weather information via radio contact. On the morning of the flight, the briefer cautions that a cold front is moving from the west toward St. Louis. Since the planned flight will reach the destination before that cold front arrives, the briefer's message raises no concern.

In the past, my flights from Marion to Fenton were nonstop and required about three hours. But today a fierce headwind pushes back. In fact, in the race with 18-wheelers below, the plane barely wins with a ground speed of 65 kts.

The strong headwind and the plane's limited fuel capacity of 22 gal force a refueling stop. I select the Burlington airport KBRL in Iowa. From that airport I could have flown directly to the Weiss airport, and all would have been well. But a different thought surfaces. It is my first mistake. I hold an STC (Supplemental Type Certificate) for the plane according to which unleaded automotive fuel, for short autogas, is a legal alternative to the lead-laden aviation fuel 100LL. Lead isn't just a health hazard but very detrimental to engine life. Since the Weiss airport doesn't offer autogas, I think, "Why not land first at the St. Charles airport KSET north of St. Louis and get autogas supplied there?" Upon landing at that airport, I find out that the terminal is deserted. A note attached to the door says that there is no service today.

A gray overcast has developed but is high enough for safe flight. Ignoring that development is mistake number 2.

The failure to refuel at the St. Charles airport irritates me and raises the stubborn thought, "I *must* get autogas before landing at the final destination." The nearby Creve Coeur airport 1Ho offers it. Hence that airport becomes the next stop: mistake number 3.

The overcast has become quite dark and the air very humid. Pilots on the ground at the Creve Coeur airport shake their heads as I take off into the gloom. But why worry? The overcast is high enough for safe VFR (Visual Flight Rules) flight,[2] and the Weiss airport is just 10 minutes away. What can possibly go wrong? Mistake number 4.

The black wall of water at the Weiss airport is just the bottom outer fringe of a huge thunderstorm invisible due to the overcast. In fact, the entire region develops closely spaced thunderstorms that afternoon. That weather development is best imagined to be a pot of water coming gradually to a boil. At that point, individual bubbles develop on the bottom of the pot, disconnect, and rise up. The water in the pot represents the atmosphere of a huge area, and each rising bubble is a thunderstorm. When a plane is trapped down low in that development, there is no escape from the thunderstorms rising everywhere, and a fatal accident is the likely outcome. That misunderstanding of thunderstorms constitutes mistake number 5.

By sheer fortune I survived all those mistakes. On the upside, that day taught important lessons about fuel management and thunderstorms.

This book recounts other close calls during 45 years of flying and what I learned from them. You may often be surprised by my ignorance as I get into difficulties. The fact is, when I learned to fly in flatland Texas and obtained the pilot license, nobody talked about the aspects covered in this book. On top, the technology supporting flight was crude and by today's standards wholly inadequate. So with little preparation I began to fly cross-country, often experienced difficulties, and learned. As my instructor Dan put it when I obtained the pilot license, "You now have permission to go out and learn all about flying." At the time, I considered that statement to be cute, but it actually is inappropriate. Pilots can learn much without experiencing close calls. If you are a pilot, this book presents an opportunity to do so.

Next are the planes I have owned and the friends who are part of the stories to come.

2
Planes and Friends

The Grumman AA1-C mentioned in the introduction was my first plane. I flew it during the 1980s and had to sell it due to an illness that ruled out piloting for a while.

The plane apparently has had at least three subsequent owners. The current one, Michael Hull, generously supplied photos. Amazingly, the plane looks just like it did in the 1980s, except that one owner added wheel pants.

Grumman AA1-C, tail number N9782U on the ground ...

...and in the air[3]

After recovery from the illness, friend and expert aircraft builder Mel Asberry and I built a Zenith 601HDS. To make it clear: He was the master and I the apprentice.

The tail number is N314LB. It references the famous number $\pi = 3.14\ldots$ and with "LB" honors the outstanding philosopher, mathematician, and scientist Gottfried Wilhelm Leibniz (1646–1716).

Mel Asberry creating one of his planes[4]

Mel has accomplished extraordinary feats in aviation: He and his wife have built nine airplanes. He has inspected more than 1,000 airplanes for airworthiness as DAR (Designated Airworthiness Representative). He is an A&P (Airframe and Powerplant) mechanic and ground instructor. He has received the EAA Tony Bingelis Award, the FAA Wright Brothers Master Pilot Award, and the Charles Taylor Master Mechanic Award. The Zenith 601HDS plane Mel and I built delivers impressive performance.

Zenith 601HDS, tail number N314LB on the ground ...[5]

...and in the air[6]

- The endurance (= max time in the air) at 65% of the rated 80 hp of the Rotax 912 engine is 4 hrs with more than 1 hr reserve. When the power setting is reduced to 55%, the endurance increases to 5 hr, and at 50% to 6 hrs, each case with 1 hr reserve. The longest flight so far has been 5 hrs.
- The ceiling (= max altitude) achieved by the plane exceeds the legal limit of 14,000 ft MSL (Mean Sea Level) for flight without supplemental oxygen.
- Takeoffs are easily accomplished even on hot days at high-elevation airports of the US.

The only drawback of the plane is slow cruise speed. When Mel and I selected the Zenith 601HDS kit, the manufacturer promised a cruise speed of 120 kts requiring just 4 gal/hr of fuel. That fuel consumption produces 65% of the rated 80 hp, and in my well-built, lightweight plane produces a much lower speed of 95 kts.

The 80 hp Rotax 912 engine has done very well. During 2,500 hrs of run time over a period of 29 years, the engine has never been opened and still performs like new: compression is perfect, oil consumption is minimal, and the oil filter shows at most a few minute metal particles during each oil change.

The run time of 2,500 hrs is far beyond the TBO (Time Between Overhauls) of 1,200 hrs specified by the manufacturer in 1994 at the time of installation. It even exceeds the current TBO of 2,000 hrs for new Rotax 912 engines and is double the specified lifetime limit of 15 years. Of course, such extended use requires careful maintenance and evaluation of all components during each annual inspection.[7]

Friends

Five friends appear in the stories. Three of them were copilots on several cross-country trips: John Barrer, Manfried Feyen, and Arie Tamir.

For several decades, John Barrer designed features of air traffic control systems at MITRE in Washington, DC. He is a flight instructor par excellence and set me straight on key problems of piloting.

John Barrer[8]

Manfried Feyen, talented industrial engineer, was born in Germany but later moved to France. He had piloting lessons in France to take off and land a plane without an instructor on board and could fly the Zenith 601HDS in an emergency.

Manfried Feyen

Arie Tamir is mathematics professor at Tel-Aviv University, Israel. He had extensive flight training at the Israeli Air Force Academy—even flying 2-engine trainer jets—and obviously could handle the Zenith 601HDS very well.

Arie Tamir (left) with the author

For many years, Arie as well as Manfried joined me on numerous flying trips lasting two or three weeks. We often took camping gear along and stayed in national parks.

The fourth friend, computer science researcher and superb pilot Jack Wybenga,[9] helped over many years with the maintenance and repairs of the two planes. He was a *bona fide* aircraft guru.

You have already met the fifth friend, Mel Asberry. Without him, the Zenith 601HDS N314LB wouldn't exist.

A few years ago, Manfried and Jack died due to cancer. I miss them.

Let's start. Part I concerns the impact of weather.

Part I

Weather

3
Headwind

May 10, 2004, 3 pm

"The GPS has failed," says Manfried as we begin crossing the Sierra Nevada west of the Mojave airport KMHV at 2,000 ft AGL (Above Ground Level).

Indeed, two dashes have replaced the ETE (Estimated Time En route) usually displayed by the Bendix KLX 135A GPS radio. The explanation for that alarming change: Groundspeed has dropped to 24 kts, and the GPS radio has concluded that we have landed!

Of course, we are still in the air, but have a headwind of 65 kts. With 24 kts groundspeed we cannot cross the mountain and reach the next refueling stop in Bakersfield, California. How did we get into this fix?

We are on a long flying trip from Dallas to Yosemite National Park. Today we started early in the morning in Grants, New Mexico, refueled in Kingman, Arizona, and now are heading west toward our final destination airport in Merced, California, a convenient gateway for Yosemite.

Near Daggett, California, forward visibility drops due to a tan-colored haze. I venture that this is Los Angeles-type pollution. But Manfried has it right, as we learn from the Daggett ASOS

(Automated Surface Observing System): This is a dust storm. We climb to 8,500 ft to escape the worst of it, but in doing so, groundspeed drops to 45 kts. I am not worried about this change since we just have a few more miles to go west. So we chug along until we reach the Lancaster, California airport KWJF.

Planned westbound route from Barstow-Daggett KDAG to Bakersfield KBFL via Lancaster KWJF and Mojave KMHV

At that waypoint we turn north for the 20 mile leg to the Mojave airport KMHV. Groundspeed goes up to 110 kts. Great! As we approach the Mojave airport, we turn northwest and climb to get over the Sierra Nevada. Groundspeed is reduced to 46 kts. I am still not worried since we have enough fuel left for the next stop.

But then comes Manfried's warning "The GPS has failed" and the realization that groundspeed has dropped to 24 kts. After briefly discussing our choices, we turn around and head back to the Mojave airport since it has three runways. On first contact, the Mojave control tower advises, "Wind 40 kts."

That's a lot of wind, but should be okay since it's coming straight down one of the runways. As I fly downwind, the tower updates, "Wind 60 kts." Ouch! This is way beyond the stall speed, so we cannot land here. In fact, even if I could get the plane on the ground, the plane would likely flip over as soon as I turn off the runway.

The tower says that the 60 kts wind is likely to continue, so waiting for a change would be futile. I realize for the first time that we may be in serious trouble.

We are bouncing around, and I cannot possibly extract the Riverside FSS radio frequency from the AOPA airports book. So I ask the Mojave tower for that information. When I call the FSS, a flight specialist responds immediately as if he has been waiting for us. I say, "We need assistance. We are over the Mojave airport, with 1 1/2 hrs of fuel remaining, and need an airport on the east side of the mountain to land."

He asks me to stand by, comes back shortly, and suggests Lancaster with 30 kts gusting to 37, and Daggett with 22 kts gusting to 29. In both cases, the wind is straight down the runway. He does not tell me, but I know, that Daggett is in the midst of a dust storm, so I rule out that airport. That leaves Lancaster.

The wind at Lancaster is far more than I have ever experienced. Manfried has stayed quite cool about the situation, but nevertheless I don't want to upset him with the confession that the next landing will be a new experience.

I dial the identifier for Lancaster into the GPS radio. Up pops the warning that this is a US Air Force airport. Back to the FSS flight specialist. He explains that he meant the General Fox airport KWJF of Lancaster. We get the ATIS (Automatic Terminal Information Service) information for that airport, contact the tower, report a 2-mile right-base position, and are cleared to land.

On final, the plane bounces up and down. We hardly seem to move forward while indicating 75 kts. As we cross the threshold, I pull the power and focus on controlling the bouncing airplane. We bump with the main gear onto the runway, are thrown up again in a strong gust, and once more bump onto the main gear. I quickly lower the nose, and we stay on the runway. As we taxi off the runway onto a taxiway that, mercifully, is slanted at 45 degrees, the tower asks for our intentions. I say, "We want to stay overnight."

We taxi to the control tower and park on its lee side. Despite that protection from the fierce wind, the plane is still shaken by wind gusts, and getting out of the cockpit without losing the hinged canopy is not easy. But with careful planning where one of us always hangs on to the canopy, we leave the cockpit and tie down. On a nearby parked Cessna, a tightly wrapped yet flapping canopy cover is scouring the plexiglass, so we don't install the canopy cover. Just walking into the wind to the terminal is difficult. Once we are in the terminal, I confess to Manfried that this was a new landing experience for me. He takes it calmly.

Later that night in the hotel, I call the Riverside FSS about the weather forecast for the next day. She says, "The earlier you can leave, the better. Winds will pick up again just as happened today, and there is another AIRMET (AIRman's METeorological information) about occasional moderate turbulence below 18,000 ft."

Once we have covered the weather forecast for the planned route, I ask, "Would you have a moment to tell me something about the situation today?" She answers, "Sure. I have two stories about this. The story of the Antelope Wind Festival, and the story of the wing walkers." I say, "How interesting! Please, go ahead and tell me."

She says, "The story about the Antelope Wind Festival is rather short. It is a festival about the wind in Lancaster." She pauses for effect, then adds, "The festival lasts each year from January 1 to December 31." I laugh while she says, "This is a joke, of course."

She continues, "But the second story about wing walkers is true. Years ago, I started out with the Flight Service when the Lancaster airport still was a Flight Service Station. When a plane would come in under high wind conditions like today, we would send out a person to the exit ramp. The plane would stop on the runway at the exit, nose still pointed into the wind. The person would go out to the plane and hang on the wingtip of the upwind wing as the plane taxied onto the ramp. This was essential for high-wing, light planes such as a Cessna 152 or 172. Otherwise, the plane would have flipped over. Then the person would walk with the plane to the tie down area. We called those guys the wing walkers."

Another Surprise

The next day we take the earliest possible shuttle to the airport, refuel the plane, and take off shortly after 7 am. The wind is already 20 kts gusting to 27. At 1,000 ft AGL, the air becomes smooth, and we climb at 70 kts groundspeed. At the Mojave airport, the mountain is already producing low-level clouds. Fortunately, they aren't of the lenticular or rotor variety that would indicate nasty turbulence. We pass under the clouds, then climb in earnest at 300 ft/min.

The rate-of-climb needle creeps up: 400 ft/min, 500 ft/min; it just keeps going. Soon we have 1,000 ft/min, then 1,500 ft/min. Finally the needle reaches the peg at 2,000 ft/min. The altimeter winds up like a toy. We reach 9,000 ft MSL, then 10,000 ft. The entire process is in completely smooth air.

The first feeling of "Wow, this is really fun" is gradually replaced by a growing worry. How high can we be pitched up? At 10,500 ft altitude, I lower the nose, first a little, then a lot. First the rate-of-climb needle does not move, but finally it leaves the peg, goes down to 1,500 ft/min, and stays there. The nose of the plane is way down, we are climbing like the blazes, and groundspeed is 100 kts. What a strange sensation!

To my relief, the rate-of-climb needle begins to drop, first slowly, then faster. The rate goes down to 1,000 ft/min, 500 ft/min, next to zero ft/min, then -500 ft/min, -1,000 ft/min, -1,500 ft/min. The needle reaches the peg at -2,000 ft/min.

While this goes on, I repeatedly retrim the plane and add power. We become nose-high, are indicating 65 kts, yet are dropping like a rock. Just like the climb, the rapid descent is in completely smooth air. It feels eerie. The descent rate slows, finally reaches zero. We are still nose-high at 65 kts in level flight, and groundspeed has dropped to 45 kts. I am reminded of the day before, or, as Yogi Berra once said, "It is déjà vu all over again."

I lower the nose and allow the plane to descend further, finally reaching 6,500 ft MSL. We are 2,000 ft AGL, the minimum recommended altitude for crossing a mountain. Yesterday we had 24 kts groundspeed, but today it is a delightful 60 kts. In 20 minutes we are on the west side of the mountain, descend into the valley, and head north. Groundspeed goes to 85 kts, and in two hours we are at the destination airport in Merced, California.

We land, tie down, get the waiting rental car, and load the camping gear. Then we are off to Yosemite National Park for a wonderful week of camping and hiking in one of America's wonderlands.

The Lesson

In the morning of the attempted flight from Grants to Merced via Kingman, the FSS briefer had laid out that there was some headwind, but that it would be manageable. I planned accordingly, and all was well until we tried to cross the Sierra Nevada.

The simple fact is: Mountain weather, in particular wind, can be full of surprises. The tremendous headwind over the Sierra Nevada and the mindboggling climb and descent the next day confirm this.

What are you to do? First, always plan on a large fuel reserve for remedial actions. Second, in the age of ADS-B (Automatic Dependent Surveillance-Broadcast) information on board, keep on checking during the flight what's happening for the remainder of the route. The next chapter has details of the second recommendation.

4
Sully's Rule

On January 15, 2009, the Airbus 320 of US Airways Flight 154 taking off from LaGuardia airport in New York City struck a flock of Canada geese and lost power on both engines. The captain, Chesley Burnett "Sully" Sullenberger III, realized that he wouldn't be able to glide to LaGuardia or the Teterboro airport, and decided to fly the plane to an emergency water landing on the Hudson River. All 155 people on board survived and were rescued.[10]

US Airways Flight 1549 as it floats on the Hudson River[11]

Later, Sullenberger emphasized the important role his copilot Jeffrey Skiles played, always stressing, "We were a team."[12] He also said, "One way of looking at this might be that for 42 years, I've been making small, regular deposits in this bank of experience, education, and training. And on January 15, the balance was sufficient so that I could make a very large withdrawal."[13]

C. B. "Sully" Sullenberger III[14]

In a subsequent interview, copilot Jeffrey Skiles mentioned a rule that Sullenberger applied during each and every flight. Every 30 minutes of the flight, no matter how long its duration, he would check conditions at the destination airport and adapt the flight plan if needed. Why did he carry out these repeated checks? The pilots of a commercial airliner flying at more than 30,000 ft don't have to worry much about en route changes of the flight plan. But at the destination the weather may force significant detours. Sullenberger wanted to prepare for any such change way in advance.

This excellent idea applies to any long-distance flight, whether high or low, or conducted under VFR (Visual Flight Rules)[15] or IFR (Instrument Flight Rules): Every 30 minutes, the pilot checks whether the remainder of the flight plan may have to be modified. For my cross-country trips, which are always conducted under VFR, Sullenberger's concept becomes the rule included below. I have named it "Sully's Rule" in deference to that extraordinary pilot.

Sully's Rule

Every 30 minutes:

- Refill the center tank by pumping fuel from the two wing tanks. When the wing tanks become dry, typically after 3 1/2 to 4 hrs, check that the destination airport can be reached within the next hour. This leaves an additional hour in reserve. If that isn't possible, change the route and go to a previously determined alternate airport.
- Check and if needed reset each and every instrument, switch, and radio of the panel. This includes the barometric air pressure of the nearest airport as reported via ADS-B, the compass direction displayed by the DG (Directional Gyro), and the frequencies dialed into the radio for upcoming communications.
- Use ADS-B to check the weather information at major airports along the remaining route and the destination airport, including wind, temperature, dewpoint, and TAF (Terminal Area Forecast).

If any of the information doesn't match expected values and results, it's time to adjust the remaining flight plan.

The scary flights described in the Introduction and Chapter 3 took place way back when I was rather inexperienced and ADS-B didn't exist. It's instructive to go back and check how Sully's Rule, which crucially depends on ADS-B, would have prevented the two scary scenarios.

In the first case, I would have seen the threatening build-up of thunderstorms west of St. Louis, abandoned any plans to get autogas, and arrived at the Weiss airport a long time before the disastrous rain and extraordinary wind shift hit. In the second case, I would have obtained winds-aloft information via ADS-B while en route, concluded that crossing the Sierra Nevada was impossible, and landed at the Lancaster airport right away.

24 SULLY'S RULE

Significant turbulence is always annoying. When it is unexpected, it can be frightening, as seen next.

5
Virga Rain

June 23, 2000, 1 pm

Manfried and I are en route to Cheyenne, Wyoming after a refueling stop in Liberal, Kansas. A large cloud, maybe 3,000 ft above us, has curtains hanging below. They signal light rain. I mention to Manfried, "We are about to get wet." Amazingly, that doesn't happen: The rain evaporates before reaching us. Another strange phenomenon: The cloud is rapidly changing shape, creating and closing wide openings.

Suddenly, all hell breaks loose. The plane zooms up at 1,000 ft/min, then descends at that rate, pitches up one wing then the other, yaws wildly to the left and then to the right, and generally behaves like a toy tossed by a child.

I grab the Y-shaped control stick with both hands and try to react fast enough so that the plane doesn't become inverted. Two minutes of this. Then we have passed the cloud. The plane still bounces around a bit, but now it is the usual reaction to thermals.

We are stunned: How did this happen?

The Lesson

We encountered *virga rain*, where a cloud produces rain that doesn't reach the ground.

Virga rain and rainbow during sunset at the South Rim of the Grand Canyon, Arizona

As rain falls from the cloud and evaporates, it removes significant amounts of heat from the air. The resulting cold air is much heavier and drops rapidly, creating microbursts extremely hazardous to aviation. That's what we experienced.

How should you handle virga rain? Two words: Stay away! If that isn't possible, fly low since the turbulence is strongest just below the altitude where the rain evaporates. In our case, we could have easily flown around the cloud with the virga rain curtains. Instead, we learned a lesson about virga rain the hard way.

Virga rain wasn't the only difficulty on the flight with Manfried from Liberal, Kansas to Cheyenne, Wyoming, as we see next.

6
Mountain Air

June 23, 2000, 2 pm

Manfried and I are on a long final for the Cheyenne airport KCYS, as ordered by the control tower. As we get down to 1,500 ft AGL, a 1,000 ft/min descent suddenly pushes us down, then a 1,000 ft/min rise yanks us up, and then another 1,000 ft/min descent flings us down again. Nevertheless I manage a reasonable landing.

The next day, the trouble continues.

June 24, 2000, 11 am

En route from Cheyenne to Cody, we reach the Boyson Reservoir near the Shoshoni Municipal airport 49U. I decide to follow the road north across the mountain ridge to the next waypoint since it seems a neat way to pass over the mountain ridge bordering the reservoir.

Route segment north of Shoshoni airport 49U

So far the air has been very smooth, and I decide on a low-level pass across the ridge. But suddenly the plane acts crazily just like the day before near Cheyenne. I swing away from the ridge and climb to 10,000 ft altitude so that we are more than 2,000 ft above the highest point. The air becomes smooth as glass.

The Lesson

In the summer, flight in the mountains past 1 pm is almost guaranteed to entail significant turbulence, as experienced during the landing at Cheyenne. There is only one remedy: Stop flying by 1 pm.

Flight across ridges and saddles can be extremely hazardous. To avoid this, stay at least 2,000 ft above the highest point. But more needs to be considered.

Never approach a ridge or saddle at 90 deg. If you encounter turbulence, any turn away from the ridge initially gets you even closer and hence into even worse turbulence. Instead, always approach a ridge at a 45 deg angle. If the air happens to be turbulent, a turn away from the ridge immediately gets you back to calmer air.

Never fly into a valley that is so narrow that it doesn't allow a U-turn. If the air becomes turbulent, you have no way to escape except for a slow and treacherous climb. In a valley allowing U-turns, fly near the upwind side of the valley. Not only is the air smoother there, but also tends to lift the plane. If you do encounter turbulence, you can readily turn around.

How can you accomplish all this? With Sully's Rule! When flying across mountains, you frequently check the surface winds at the airports ahead and the winds aloft. With that information, you can avoid turbulence and nasty downdrafts and at times even exploit the winds to lift the plane.

For a brief yet comprehensive summary of additional rules and suggestions about mountain flying, see the Notes.[16]

Additional Recommendation

Before you head to the mountains for the first time, determine the performance of your aircraft by climbing as high as the plane will take you with the anticipated payload. Watch the climb rate deteriorate and declare the ceiling to be the altitude when the climb rate has dropped to 150 ft/min. This is higher than the 50 ft/min used by manufacturers to establish the plane's performance, but provides a more realistic estimate for the performance in the mountains. If you like precision in your calculations, you may use density altitude for the tests and the subsequent application of the information.

The pilot of a small plane I met years later at the Riverton, Wyoming airport KRIW wished he had done so. He was based in Michigan, and this was his first trip to the mountains. It turned out that his plane wouldn't climb past 9,000 ft MSL. He asked me how he could proceed westwards from Riverton. There was no reasonable choice. For example, the nearby Togwotee pass west of Riverton crests at 9658 ft MSL. I suggested that he repitch the ground-adjustable propeller for safe flight in high terrain. He was loath to do so since he didn't have the tools, and insisted on continuing. I have no idea what he eventually decided.

Chapter 3 discussed a headwind that prevented Manfried and me from reaching the destination airport. The next chapter looks at a seemingly similar, but actually quite different case.

7
More on Headwind

July 8, 2023, 9:00 am

As I take off from the Greeley, Colorado airport KGXY and climb to 2,000 ft AGL, the westerly wind of 25 kts at departure increases to 35 kts. That makes it a strong headwind since the destination airport KRIW in Riverton, Wyoming is northwest of Greeley.

Worse yet, the terrain rises, and the headwind creates a significant downdraft. I counter it by raising the nose. The total effect of headwind and downdraft: The plane slows down to an anemic groundspeed of 40 kts, which means that I cannot cover the 230 nm miles to the destination airport before mountain-induced afternoon turbulence sets in.

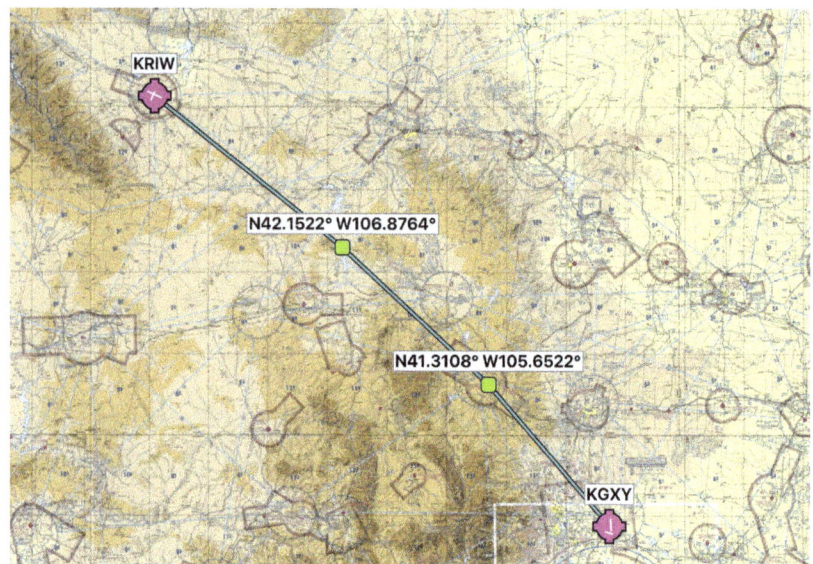

Route from Greeley KGXY, Colorado to Riverton KRIW, Wyoming

Sounds like the situation of Chapter 3, where a severe headwind stymied Manfried and me, doesn't it?

Yes, it has the looks of it. But this is many years later, in 2023. I have ADS-B available, use Sully's Rule, and have much more experience flying across mountains.

The Solution

The key is the behavior of wind in valleys. Whenever a valley doesn't line up exactly with the direction of the wind—in this case, not east-west—the wind tumbles into the valley on the downwind slope and produces a strong downdraft. In contrast, on the upwind slope, that turbulent air gets compressed and rises, thus creating a smooth updraft.

The solution to downdrafts in the mountains then is: Find valleys that satisfy two conditions: They should be reasonably close to the route, and they shouldn't line up exactly with the wind direction.

Then fly on the updraft side of these valleys. You not only avoid turbulence and downdrafts but can convert the climb induced by the updraft into additional speed.

How do you identify these valleys? For nearby ones, just look around. But to discover suitable valleys in the distance, small roads are most helpful. When engineers designed them, they balanced construction costs versus directness of route. For small roads the traffic is limited and directness isn't so important, and the construction costs become the major concern. As a result, small roads in the mountains often proceed along the bottoms of valleys. If the wind direction doesn't match the direction of such a road, you can stay on the upside of the valley to get an updraft.

I use this insight extensively during the trip to Riverton. Here is one case.

Northern part of Seminoe Reservoir, Wyoming

The photo shows the northern portion of the Seminoe Reservoir. The map below displays the entire reservoir. The planned route is drawn in dark green. It proceeds roughly parallel to the easterly shore of the reservoir.

34 MORE ON HEADWIND

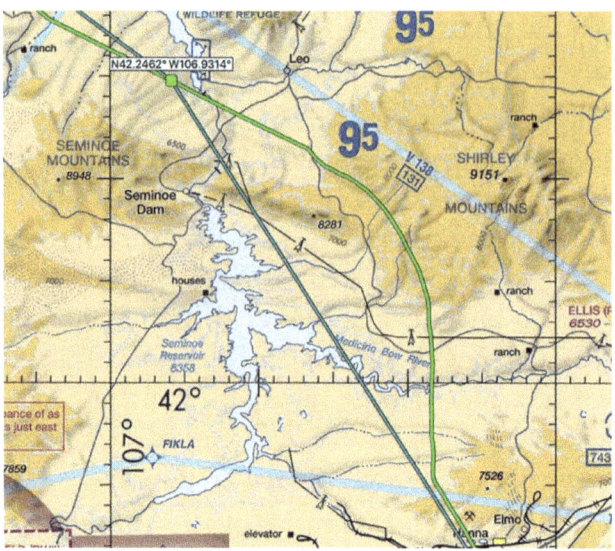

Planned (= dark green) and actual (= light green) route near the Seminoe Reservoir

Near the northern end of the reservoir, the planned route crosses a mountain range with the highest terrain to the west. Surely the strong westerly winds tumbling down from the peaks are producing significant turbulence and a strong downdraft for that part of the route.

The actual route is displayed in light green. See the small wiggly road to the left of the actual route? It indicates a valley, with the road at the bottom, and is the reason for the detour. Proceeding on the upwind side of that valley as shown, a strong updraft lifts the plane in a few minutes more than 3,000 ft. That terrific effect makes the remainder of the flight to Riverton easy.

Let's move to a different weather topic: clouds. They are very important since VFR flight must stay away from them.

8
Sudden Clouds

June 24, 2000, 7 am

Before takeoff from the airport KCYS in Cheyenne, Wyoming for the flight to the Cody, Wyoming airport KCOD, I tell Manfried, "This will be a simple flight in terrific weather across scenic mountains." Minutes later that claim proves to be false. When I contact the FSS to activate the flight plan, the flight specialist says, "There is a weather update for your route to Cody. The portion from Douglas to Casper is IFR."

The weather in Cheyenne had been perfect at sunrise: Not a cloud in sight and visibility unlimited. Pilots jokingly call such weather *severe clear*. Two hours earlier, the weather forecast by the FSS had promised this for the entire flight to Cody. How is it possible that a portion of the route suddenly switches from perfect VFR to IFR when the forecast obtained just two hours ago predicted ideal weather for the entire route? And what shall we do now?

The thought pops up, "This is just ground fog. I should ignore this and simply climb over it when we reach Douglas." The technical term is VFR over-the-top. One look at the mountains to the left squashes that idea. Dense cloud layers roll down the sloping terrain like carpet coming off a production line. Imagine what the route from Douglas to Caspar must look like!

That sobering thought brings me back to reality. There is no way we can fly from Douglas to Caspar using VFR over-the-top. Fortunately, there is a safe alternative: Cloudless terrain lies to the east. Consulting the map, I opt for the Wheatland airport KEAN to stop, refuel, and wait for better weather. Soon other pilots land there too, all stymied by the sudden shift in weather.

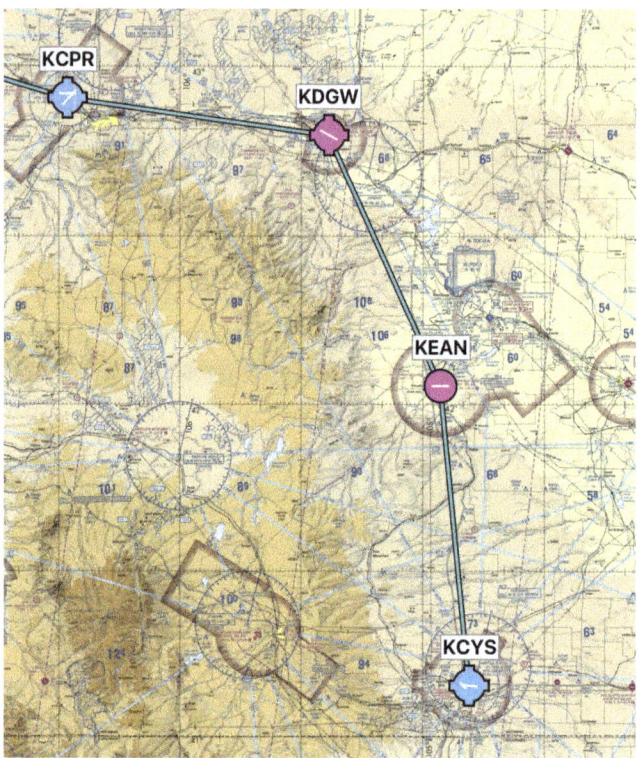

Detour from Cheyenne KCYS to Wheatland KEAN, then via Douglas KDGW to Cody KCPR

The weather improves by 9 am. We take off and reach Cody around noon. That easy resolution masks major blunders on my part: I didn't anticipate the sudden appearance of clouds and didn't plan a suitable route alternative for that scary event. I even considered VFR over-the-top, a horrible choice in that weather!

What would have been the right way to handle the situation?

The Lesson

Two conditions together virtually assure low level clouds shortly after sunrise: First, temperature and dewpoint are within 5 deg F (2.5 deg C). Second, no wind. When this happens, you should delay the takeoff and wait until the sun has risen and warmed up the ground sufficiently to increase the temperature/dewpoint gap. Even then, be prepared that you may have to go to an alternate airport first and wait for improving weather.

During the rapid growth of clouds near Cheyenne, VFR over-the-top would have been a terrible decision. The next chapter lays out a complicated set of conditions under which I consider that choice to be justified.

9
Above the Clouds

"VFR flight above a cloud layer is dangerous and should never be done," many pilots counsel. In support, they cite statistics about such flights with fatal outcomes. For many years I agreed with that assessment, but then analyzed the situation and came to a different conclusion: When such a flight is carefully prepared and conducted, it becomes quite safe. One thing is certain: The unaccustomed view of cloud formations from above makes such flight exhilarating.

Flying above an overcast

To start, let's look at the cloud conditions for VFR flight.

VFR Cloud Conditions

Below 10,000 ft MSL, VFR flight must be at least 500 ft below clouds or 1,000 ft above clouds, and horizontally must be at least 2,000 ft away from clouds.

Above 10,000 ft MSL, two of the conditions are stricter: The 500 ft condition for flight below clouds is increased to 1,000 ft, and the horizontal distance condition becomes 1 mile.

The conditions are relaxed close to the ground: Up to 1,200 ft AGL, VFR flight only needs to stay clear of clouds. At airports with surrounding magenta boundary, that limit is lowered to 700 ft AGL.[17]

The above summary is incomplete since it ignores conditions for night flight, which I never dare to do, and since it skips over some subtleties as well as certain visibility conditions. But the summary suffices for the present discussion.

Also needed are the concepts of cloud types and ceiling.

Cloud Types and Ceiling

Clouds are characterized as follows: There are few clouds, or the clouds are scattered, or the cloud cover is broken, or there is an overcast. Percentage figures for cloud coverage characterize these types of cloud covers. In flight these numbers are not so useful. A practical guide is that a scattered layer consists of individual clouds that are separated by large holes allowing easy penetration under VFR conditions. When such penetration is no longer possible but holes are still present, the clouds constitute a broken layer. When there are no holes at all, it's an overcast.

Cloud layers, of whatever type, are always given with the altitude of the bottom of the layer, measured in feet AGL.

Scattered layer, Rocky Mountain National Park

Broken layer, Rocky Mountain National Park

Consequences of the Cloud Rules

Suppose you have the following situation. You take off under VFR conditions and later encounter an extensive area of clouds so low that you cannot fly below them. You have two choices: You land

and wait for the clouds to lift, or you climb above the clouds and continue.

If you go for the first choice, you may have to wait for several hours, and thus may not be able to make it to the planned destination the same day.

In case of the second choice, the cloud deck may extend beyond your destination. Thus, when you reach that point, there is no safe way to descend. Worse yet, while you are flying above the cloud deck, the clouds may rise higher and higher. Accordingly, you must climb to stay 1,000 ft above the clouds. In the nastiest case, you get to the point where the plane cannot outclimb the clouds or where you reach the limit of 14,000 ft MSL for flight without supplemental oxygen. Terrifying situations indeed.

Here is another situation where flying above a cloud layer seems attractive but ultimately is the wrong choice: You could fly below a cloud deck, but the ride would be bumpy or would encounter a strong headwind. In contrast, above the clouds, the flight would be completely smooth with a tailwind. Lured by these attractive flying conditions, you begin flight above the cloud deck, but later run into one of the above problems.

Flying on top of a cloud deck under VFR conditions is called "VFR over-the-top." How can you safely conduct such flight and avoid terrifying scenarios?

VFR Over-the-Top

VFR over-the-top can be simple. For example, near the coast, there often is a low-level cloud bank. You rise above the cloud bank, cross it, and descend again once you are over open water. Similar situations can occur when you cross a mountain range, or when you face a well-defined cluster of clouds near a lake.

Cloud bank, Grand Teton National Park

The situation is much more complicated when the cloud cover extends beyond the visible horizon. Let's see how you can safely conduct VFR over-the-top in that setting.

Equipment

You should consider VFR over-the-top for extended distances only if your plane has the equipment described in Appendix A: ADS-B In and Out, a flight management system such as the Garmin Pilot or Foreflight, a GPS-based backup system, and an autopilot.

The autopilot not only keeps you reliably on the intended route, but also frees you for the extensive checking described in a moment. The ADS-B information provides timely weather data for all airports with weather reporting equipment. The data include TAFs for major airports. The forecasts give you weather predictions well beyond the duration of the flight. ADS-B also supplies a radar image of rain and thunderstorms. Needless to say, I assume here that the plane has performed flawlessly for an extended period before you consider VFR over-the-top.

There are two decision processes for VFR over-the-top. First, should you begin it? Second, should you continue?

Decision to Start VFR Over-the-Top

First, you must have clear sky or a scattered layer to climb through for VFR over-the-top. This is easily ascertained.

Second, you must be sure that the broken or overcast layer does not rise beyond the capability of the airplane or the legal limit for flight without supplemental oxygen. Here is our much more stringent condition: You must be sure that the layer rises at most by a trivial amount. To ascertain this, you look at all reported weather data of airports along or near the planned route. This includes checking of all available TAF data applying during the entire flight.

Third, you must be sure that prior to the destination, you have clear sky or at most a scattered layer to descend through. TAF data are crucial here. All airports near the destination with TAF data must confirm this condition.

Fourth, do not consider VFR over-the-top where the terrain below the clouds is hostile, such as mountains. If an equipment failure forces an emergency descent through the clouds, you won't have enough time to look for a reasonable landing area.

Decision to Continue VFR Over-the-Top

While proceeding above the clouds, keep on looking up the weather conditions for the airports en route and at the destination. This includes checking TAF data: They are unlikely to change, but checking them time and again is a good idea. Do this at least every 10-15 minutes. If things deteriorate, look for an alternative, including turning back. It is essential that your plane has significant endurance that makes these safe alternatives possible.

Special Aspects

The airports along the route and at the destination may report a very high layer, say at 25,000 ft AGL, that may be broken or overcast. You can safely ignore that layer. But if such a layer is not far above your intended altitude, you effectively would be sandwiched between them. Do NOT attempt any such flight since you may be squeezed from the bottom and top to the point where VFR is no longer possible.

Never attempt VFR over-the-top when the broken or solid cloud layer below starts at several thousand ft AGL and is also rather thick. Such clouds can easily puff up and trap you on top. The cloud layer below should be at most 2,000 ft AGL and rather thin and uniform, say no more than 1,000 ft thick.

Never attempt VFR over-the-top when significant rain is predicted or occurring, or when thunderstorms are developing or in progress.

When flying several thousand feet above an overcast, dark areas may appear in the distant cloud cover.

Apparent holes in cloud cover near the horizon

These areas seem to indicate that holes are opening up due to the warming effect of the sun and descent will soon be possible. But don't be fooled. A scattered layer of clouds far above, say at 25,000 ft, can produce these dark areas.

If you fly toward such supposed holes, they always disappear as you get closer. To decide if holes are really opening up in the distance, look at the weather reports of airports in that area.

Conclusion

VFR over-the-top must be carefully managed to be a safe operation. At times, it can be the only option. For example, in a recent flight from Dallas to Albuquerque, it was the only way to transition across Northwest Texas and Eastern New Mexico, an area that during spring and fall loves thin overcasts at 500 ft AGL that sometimes last all day.

Flight below clouds demands careful management just like flight above. The next chapter recounts how once I almost got trapped by low-level clouds.

10

Below the Clouds

May 16, 2010, 10 am

Thirty miles from my next refueling stop at the airport KPVW in Plainview, Texas, clouds appear at 400 ft AGL contrary to predictions. First a few clouds, then scattered. Finally, 15 miles from Plainview, the layer at 400 ft becomes broken, meaning that VFR flight to the airport is impossible. The development is contrary to the weather forecast, which had promised clouds high enough for VFR flight. Now what?

It's a case of my overconfidence in weather forecasts. Worse yet, I hadn't even prepared to go to an alternate airport. On the plus side, there is plenty of fuel left. I consult the sectional map, select the nearby airport 3F6 in Richards, Texas and land there. Then I find out that the airport doesn't supply fuel. At that point I realize how badly I have prepared for this flight.

Scrambling, I fly north to the Childress airport KCDS, refuel, and take off for the final destination of the day: the airport KAEG in Albuquerque, New Mexico.

It isn't a day to be proud of and forces me to rethink planning of long-distance flights to account for capricious weather. The result is the following recommendation.

Forecast Evaluation

For 2-3 days ahead of a planned long-distance flight, check in the morning what weather is predicted for that day. Later in the day, verify how much of that prediction did and didn't come true. Also determine how much the forecast for the departure date varies.

Two charts are the key tools for this process.

- Forecast weather: The Prog Chart produced by the National Weather Service of the NOAA (National Oceanic and Atmospheric Administration) supplies a comprehensive forecast for the morning, noon, afternoon, and midnight of the current day; the morning and afternoon of the next day; and the morning of the next six days. Search the Internet using "prog chart" or use the url https://aviationweather.gov/gfa/#progchart.

- Actual weather: The radar image of the current weather is available from many sources on the Internet. We use the AOPA website https://www.aopa.org/weather/.

Here is an example Prog Chart.

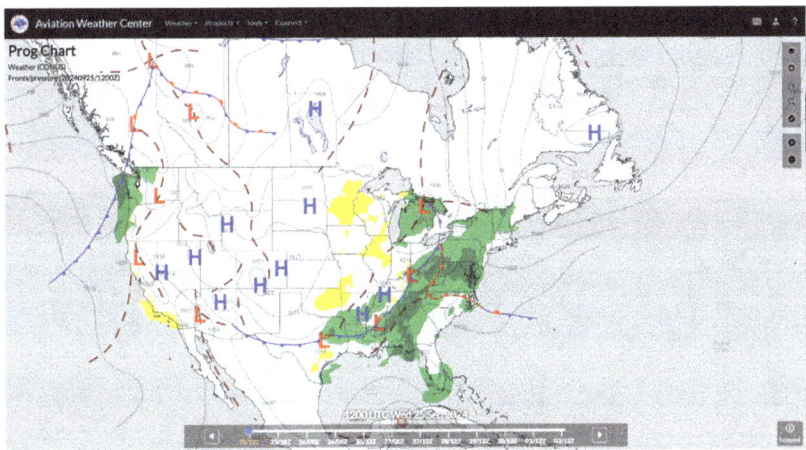

Prog Chart of Aviation Weather Center

The Aviation Weather Center has an almost overwhelming amount of aviation data beyond the Prog Chart. The AOPA weather website provides key parts of that material in simplified and easy-to-interpret form.

Forecast Evaluation produces an insight that you can't possibly get by just looking at the weather forecast on the morning of the day of departure. Indeed, suppose that Forecast Evaluation has determined major discrepancies between forecast and actual weather during the past few days. You then can work out suitable alternate routes that take these differences into account while you develop the flight plan.

―――――――

Knowledge about the precision of weather forecasts obtained by Forecast Evaluation enhances the use of forecasts in Sully's Rule. The next two chapters demonstrate this.

11

Thunderstorms

August 15, 2002, 2 pm

Arie and I are approaching the Orlando, Florida area on a trip to the Florida Keys. The next stop will be the airport KMLB in Melbourne just southeast of Orlando. Rain clouds have spawned a thunderstorm cell that has come ashore from the east. It's quite contrary to the forecast by the FSS flight specialist, who had predicted light rain. Nevertheless, I conclude that there is no problem: We will bypass the cell on the east side near the shore and, presto, be in Melbourne. As we get closer, I notice that the cell, now to our right, has a cousin that has come ashore south of us, right in our path. Again, no problem. We will move a bit more east to bypass that cell as well, and end up in Melbourne. But wait, we are so far south that going even two miles east would put us into Restricted Area R-2934 of the Kennedy Space Center, which we are not allowed to enter. Now what?

The cells are small and nothing like the West Texas monsters where getting closer than 10 miles invites havoc. They are separated by 15 miles, and there is no turbulence. We decide to go west between them, swing around the second cell, turn east again, and reach Melbourne. In bursts, moderate rain splashes on the canopy. After a few minutes, we come out of the rain into sunshine and see the Melbourne airport 10 miles ahead of us.

A clever solution? At the time I thought so. But actually I committed a major blunder. The solution—passing around the front of the second thunderstorm—is a no-no since one may become embedded in the violent weather. In this case we were lucky: We moved fast enough that the thunderstorm didn't catch up with us.

The safe solution: We should have circled for a while, say 15-30 min, to let the second thunderstorm pass, then fly behind it to Melbourne. There was plenty of fuel on board to do so.

Types of Thunderstorms

The experience gave me a first taste of Florida thunderstorms. They can appear quickly, but fortunately are often of moderate size. That's different from thunderstorms of the Midwest, which typically build gradually but may grow to significant size. But even the Midwest cases are minor cousins of monster West Texas thunderstorms, which can be huge and very violent. Finally, thunderstorms in the Rocky Mountains produce extraordinary turbulence due to mountain effects.

This means: There is no such thing as a typical thunderstorm in the US. You can only cope with thunderstorms successfully if you anticipate behavior typical for the region. But there is one ironclad rule: You must stay at least 10-15 miles away from them.

The two thunderstorms in Florida happened to Arie and me when small planes couldn't have weather radar information on board. Today, with ADS-B or some other system supplying that information, you can cope with thunderstorms safely and effectively. The two methods mentioned in earlier chapters, Forecast Evaluation and Sully's Rule, are key tools. Here are details.

Planning for Thunderstorms

Whenever you are planning a long-distance trip where thunderstorms might pop up, start looking at weather forecasts 2-3 days before the flight. Each day, check to what extent the forecast was correct. As part of that Forecast Evaluation, examine where thunderstorms were to come up, how they were to move, and what actually took place. As part of Sully's Rule, track during the trip itself how thunderstorms rise, move, and dissipate. The radar picture and the weather reports at airports along the route supply the key information. In particular, lightning information added to the radar image supplies good estimates for the location of thunderstorms.

Of course, looking out of the canopy is a good idea, too! You then match thunderstorm cells implied by lightning on the map with actual thunderstorms you see. Here is a typical view.

Typical thunderstorm over a field[18]

When you compare the radar images with the outside view, you must take into account that the radar information typically is 15-20 min old.

The next chapter covers two example situations.

12

Managing Thunderstorms

July 7, 2024, 9 am

Here is a seemingly daring route from Dallas to the Canadian, Texas airport KHHF. The Childress, Texas airport KCDS is a waypoint.

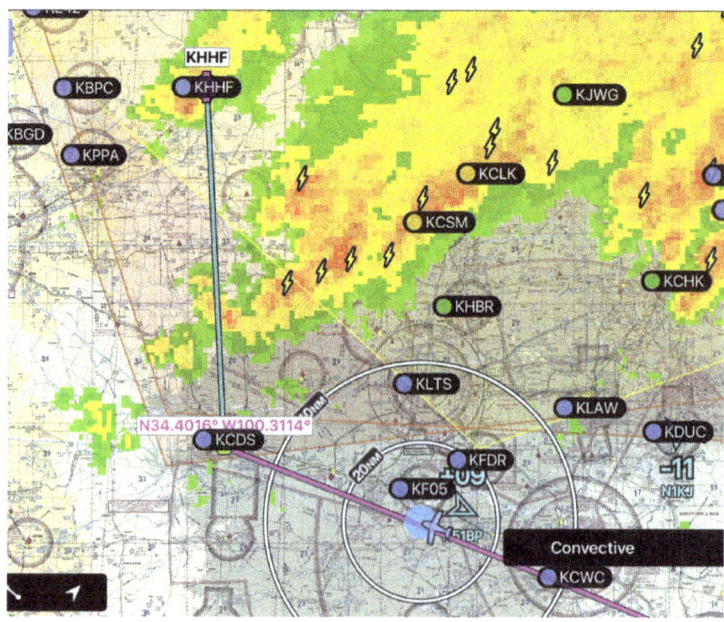

Route to Canadian KHHF with Childress KCDS as waypoint

MANAGING THUNDERSTORMS 53

The green segment of the route cuts through yellow/orange/red areas determined by radar, indicating significant rain. The destination aiport KHHF is embedded in such an area. Lightning from thunderstorms is occurring nearby. All of this is embedded in a brown-shaded area produced by a SIGMET (SIGnificant ME-Teorological information). It warns of possible thunderstorms in that region but indirectly also indicates where such nasty weather should *not* occur. The latter information is useful for planning escape routes.

Why did I dare select the displayed route? Over two days, Forecast Evaluation had determined that predictions of thunderstorms moving from southwest to northeast were quite reliable. This turns out to be correct on the flight day as well: Thunderstorms blanketing some portion of the route always have moved northeast by the time I get there. Indeed, I never encounter rain let alone a thunderstorm, and clouds are high enough for safe VFR flight.

The weather picture on the next page shows the second flight on the same day, from Canadian KHHF to Goodland, Kansas KGLD. It looks very similar, doesn't it?

There is a caution sign, though. While the blue dot of the destination airport KGLD tells that VFR conditions prevail there, the nearby airport KITR in Burlington, Colorado has a yellow dot, indicating IFR. A short time later, that worry is eliminated as the rain and thunderstorms proceed to the east as predicted and Burlington is upgraded to marginal VFR.

But as I get closer to the two airports, new bad weather that wasn't predicted moves in from the northwest, and both Burlington and Goodland become IFR. It's time to move to the alternate plan. It relies on the fact that clear skies prevail west of the route. Detouring to the west, I bypass Burlington and Goodland and eventually land at the Akron, Colorado airport KAKO. Time in the air for the trip from Canadian KHHF to Akron KAKO is 4.8 hrs, significantly more than the 3.5 hr allotted for the originally planned route to Goodland.

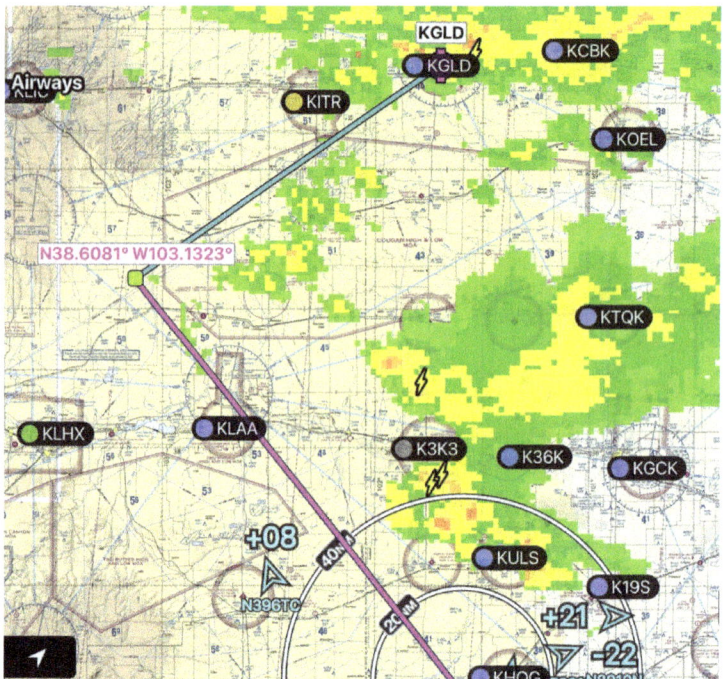

Route from Canadian KHHF to Goodland KGLD

Conclusions

Safe long-distance flight in challenging weather requires careful Forecast Evaluation, route planning that includes safe alternatives, continuous monitoring of weather as part of Sully's Rule, ample fuel for extended endurance, and—this may surprise you—an autopilot. The latter device relieves you of managing course direction and thus frees you up for detailed analysis of alternatives. That's a major reason we consider an autopilot essential for long-distance flying. In addition, flying with an autopilot for hours is far less tiring, and you arrive at the destination alert and relaxed for a safe arrival process.

Finally, the two cases look deceptively similar when one just examines the maps with radar information, don't they? But with Forecast Evaluation and Sully's Rule it becomes obvious that the first route is safe, while the second one must be abandoned and replaced by a suitable alternative.

High temperatures at an airport can create havoc with takeoff performance. We learned this the hard way, as we see next.

13
Hot Air on the Ground

July 27, 1984, 1 pm

The engine of the Grumman AA1-C feels sluggish as I apply full power for takeoff at the Tucumcari airport KTTC in New Mexico. The plane accelerates rather slowly and only reaches lift-off speed when we have eaten up most of the 7,100 ft runway. I am shocked by this poor performance as is Arie.

Once we are en route to our destination, the Santa Fe airport KSAF in New Mexico, I reduce power to 75% and pull the mixture control slowly for maximum rpm. We reach Santa Fe midafternoon and tie down.

Arie points to the runway and says, "Look at the Cessna 152 taking off." The plane accelerates very slowly, lifts off after using most of the runway, then climbs at a feeble rate. Seeing that performance of a plane with much bigger wings than our Grumman, we agree that on the day of our return trip to Dallas we will go to the airport before sunrise and lift off while the air is still cool. We implement that plan several days later, lift off after a reasonable takeoff run, and establish a solid climb for the flight back home.

Why was the takeoff performance in Tucumcari so poor? I thought I was doing everything right, using the following rules I had learned for airplanes with a carbureted engine.

- When flying at less than 75% of rated power, pull the mixture control until highest rpm with smooth running engine is attained.
- When flying above 5,000 ft MSL (Mean Sea Level), do so even if the power setting is above 75%.

Since the Tucumcari airport is at 4,065 ft MSL, the second rule did not apply, and the takeoff had to be conducted with full-rich mixture. Why didn't this work?

"The second rule is wrong," explained guru Jack at my home airport when I told him about the marginal, indeed dangerous takeoff in Tucumcari. He added, "You were missing something else, too." Indeed, I had committed two major errors.

The Lesson

Here is the correct information: A revised second rule plus a third rule for takeoff. Here is the new second rule.

- When flying above 5,000 ft *density altitude*, pull the mixture control for maximum rpm with smooth running engine.

See the small difference? "MSL" has been replaced by "density altitude."

When Arie and I had landed at Tucumcari, the airport reported 95 deg F temperature. That heat converted the airport elevation of 4,065 MSL to roughly 7,000 ft density altitude, which was well above the threshold value of 5,000 ft of the revised second rule. Hence I should have leaned the fuel mixture.

But more went wrong, as Jack pointed out. I should have used the following third rule for takeoff.

- If density altitude is above 5,000 ft, conduct the takeoff as follows. Step on the brakes, apply full power, then gradually pull the mixture control until maximum rpm with smooth running engine is attained. At that point release the brakes and start the takeoff run.

If I had applied these rules, the takeoff in Tucumcari would have used much less runway and had been followed by a solid climb.

How can you easily get the density altitude for application of these rules? If you have a fancy glass panel, you may have the value readily displayed. If you are not so lucky—like me—here is a simple process that supplies a reasonably accurate value and applies whether you are on the ground or in the air.

Computing Density Altitude

Divide the temperature T in Fahrenheit by 20. Divide the displayed altitude A by $4,000$. Add the two numbers, then subtract 3, getting a number E. Multiply E by $1,000$, and add it to the altitude A to get the density altitude D.

The computation may sound complicated, but when you see it in a formula, it becomes easy to understand and remember.

$E = T/20 + A/4,000 - 3$

$D = A + 1,000 * E$

In fact, the computations are easy to do when you use rounded values for $T/20$ and $A/4,000$.

In the Tucumcari case, we had $T = 95$ and $A = 4,065$, and hence $E = 95/20 + 4,065/4,000 - 3$. Rounding the division results where $95/20 \approx 5$ and $4,065/4,000 \approx 1$, we get the approximate result $E = 5 + 1 - 3 = 3$. Hence we add 3,000 ft to the elevation 4,091 ft of the Tucumcari airport to obtain 7,000 ft as approximate value for the density altitude.[19]

Simplified Rotax Procedure

The carburetors of the Rotax engine automatically adjust for altitude due to an ingenious design. This simplifies the takeoff when density altitude is high: Step on the brakes and apply full power. When the engine has achieved maximum rpm, release the brakes and start the takeoff run.

Engine Performance

Density altitude is directly related to engine performance. The simple formula: Each 1,000 ft of density altitude reduces engine horsepower by 3.5%. For example, at 7,000 ft density altitude, the max engine output is about $100 - (7 \cdot 3.5) \approx 75\%$ of rated horsepower. That number is essential for predicting the length of the takeoff run.

If the manufacturer of your plane or the kit used to build the plane does not supply takeoff data for various scenarios of payload and density altitude, you can easily build that information by recording the length of the takeoff run for various situations. With that data in hand, try to define a simplified rule. Here is the takeoff rule for our plane derived that way. It applies to takeoff from hard-surface runways. That's appropriate since we never consider any other kind of runway except for emergency landings.

Takeoff rule for Zenith 601HDS N314LB

Regardless of the payload, the takeoff run requires at most 2,000 ft of runway when density altitude is below 6,000 ft, and never uses more than 3,000 ft up to 10,000 ft density altitude.

The next chapter describes a case of hot air up-high.

14
Hot Air Up-High

August 15 , 2006, 6:00 am

Arie warns, "Oil and coolant temperature are at the upper end of the green arc" as we are climbing out through 1,500 ft AGL after takeoff from the Bullhead City airport KIFP in Arizona. The airport is directly across the Colorado River from Laughlin, a gambling city in the southernmost tip of Nevada, where we have spent the night. I think, "How is this possible? It's still before sunrise, and we just had 92 deg F at takeoff!"

Worse yet, the oil and coolant temperatures are now moving beyond the green arcs. We are in untested performance territory of our Zenith 601HDS.

Yet, we must continue to climb if we are to get out of the Colorado Valley. How on earth did we get into this predicament? In a nutshell: We flew into Bullhead City without any understanding of weather in the desert.

The Trip

The day before: We fly from Grants, New Mexico to Bullhead City via Flagstaff, Arizona. On the 200 nm leg from Flagstaff to Bullhead

City, I must reset the altimeter three times, for a total correction of 2 inches. In terms of altitude: If I had not done so, the altimeter reading would have become low by 2,000 ft, a critical error.

As we descend into the Colorado Valley for Bullhead City, the temperature indicated by the OAT (Outside Air Temperature) gauge climbs. We had seen this on previous trips. But now the temperature goes up past anything we encountered before. By the time we land, the gauge shows 115 deg F.

I do not think about the implications, and we refill the tanks right to the filler necks, take the small ferry across the Colorado River from Bullhead City to Laughlin, and check into one of the hotels. Arie has spent time in the Sinai Desert, courtesy of the Israeli Army. As we walk in late afternoon to the Flamingo Hotel for their outstanding buffet, he says that he never felt that hot in the Sinai. It turns out to be 125 deg F.

I am beginning to worry about the next day's departure. During the night, I don't get much sleep and work out a detailed plan for the takeoff. Too bad that the tanks are full. We could take off with 50 lbs less fuel and still make it to our next stop, Cedar City, Utah. But there is nothing to be done about this major mistake.

The main idea is to get to the airport before sunrise, take off as soon as the FBO opens the gate, and then climb out while staying near the airport. Optimal cooling during climbout with gross weight occurs at 70 kts and 4,700 rpm. This is cruise power output and results in an anemic climb of 50-100 fpm (feet per minute). In all our travels, that speed and power setting got us out of the hottest places. If, to the contrary, the oil or coolant temperature becomes too high, we will pull power, land again, and go to plan B yet to be devised.

To our pleasant surprise, the air feels almost cool before sunrise as we take the ferry across the Colorado River from Laughlin to the airport. Luck has it that an airport employee shows up early at 5:30 am. We plead our case. Kindly, he opens the gate for us. During the preflight inspection I see that the temperature gauge shows 92 deg F. I think, "Great, there wasn't really any reason to worry."

I start the engine, taxi to the end of the runway, check engine performance during the run up, and off we go on runway 36. The temperature gauge still shows 92 deg F.

As we turn left over the Colorado River and head south, passing over the hotels and casinos of Laughlin, the OAT climbs to 95 deg F. We saw that temperature on another trip in California's Central Valley and climbed out easily. So still no need to worry.

A minute later, I look again at the temperature gauge. To my amazement, the temperature has increased to 96 deg F. Then it moves up even higher: 97, 98, 99, and finally 100 deg F, reached at 1,200 ft AGL. Regardless, we must continue the climb to at least 2,000 ft AGL so that we can legally fly over the Davis Dam and proceed north across the recreational area of the Colorado River toward the Hoover Dam.

I burst out, "This is awful," but in much stronger language. It surprises Arie, who has monitored the engine gauges but not the OAT. I point to the OAT gauge, and he realizes that the reading isn't normal. I ask him to watch the oil and coolant temperatures continuously and give me readouts as the numbers change. In turn, I hold the airspeed precisely at 70 kts and try to get extra lift from the slight wind moving up the sloping terrain.

It feels very hot. Of course, the air vents! They are open and blasting us with 100 deg F air. We close the vents. This is a first: closing the air vents to stay cool!

The green arc of the oil temperature gauge ends at 230 deg F, in agreement with the Rotax manual. But Rotax allows oil temperatures up to 285 deg F for a short time. Since I do not know how long we will be forced to fly in the hot air, I reduce that limit to 250 deg F to be on the safe side. The green arc of the coolant temperature ends at 220 deg F. This was my decision when I marked the gauge. In almost 1,000 hrs of flying, I have seen at most 215 deg F. A 50/50 coolant with 12 lbs pressure cap boils somewhere between 240 and 250 deg F at sea level. The value declines slowly with altitude. To be safe, the checklist defines a limit of 230 deg F. So these are the critical values: 250 deg F oil and 230 deg F coolant. If either temperature goes beyond, we must break off the climb and return to the airport.

Arie reads out slowly rising oil and coolant temperatures. They are marching steadily toward the limits, even though we are climbing at less than 100 fpm. A few miles south of the airport, I begin to fly wide circles. The temperatures keep going up, then finally stabilize exactly at the limits of 250 deg F oil and 230 deg F coolant. All the while, the OAT stays at 100 deg F.

Reaching 1,900 ft AGL, I turn north and cross the Davis Dam above 2,000 ft AGL as required. As we achieve 2,500 ft AGL, the OAT drops to 99 deg F. Never before have I liked that temperature so much. Even better, after a while it's 98 deg F. Ever so slowly, we get further OAT decreases as we gain altitude. Once the OAT has dropped to 95 deg F, the oil and coolant temperatures are at the top end of the green arcs.

As the OAT drops further, I add power while keeping both oil and coolant temperatures in the green. The plane climbs steadily into the glorious morning. Arie and I start to relax and enjoy the view of Lake Mead and the surrounding dark volcanic mountains.

Analysis

The rapid climb of air pressure as we approached Bullhead City the day before implied that we were flying toward the center of a huge high pressure system. This meant that air movement would be almost nonexistent.

Second, the temperature at Bullhead City dropped overnight at ground level since the heat of the ground radiated away due to the dry desert air. But just a bit above the ground, from 1,000 ft AGL upward, there was little such radiation effect. These two events produced the temperatures on the morning of departure: a reasonable value at ground level, and a rapidly rising temperature after takeoff.

The Lesson

If a high pressure system is centered near a destination airport, there is virtually no air movement, and a high temperature at *ground level* on the day before the flight is a good predictor that the temperature *aloft* on the day of the flight will be high, too. A more precise forecast is the temperature aloft if you come in for a landing in the afternoon: You can expect the same temperature aloft when you depart the next day in the morning. This phenomenon isn't restricted to desert areas. We have experienced it in the summer as far north as Montana.

Additional Recommendation

If you have not done so, establish maximum temperatures for engine oil and—in the case of a water-cooled engine such as Rotax—water coolant, that you will never exceed. If during preflight planning you determine that air temperatures may possibly push the plane beyond those maximum values, establish a plan B of alternate routes and airports.

If a maximum value for oil or coolant is marginal, try to change the plane's cooling system to improve it. We did so with an electric cooling fan for the oil radiator.[20] As for the water temperature, a detailed analysis showed that 240 deg F, and not 230 deg F as assumed during the above flight, is a conservative maximum value.[21] Accordingly, we have not modified the coolant system.

Let's look at some fundamentals of flying.

Part II

Fundamentals

15
Traffic Pattern

September 30, 1984, 1 pm

"Don't do that!" shouts John as I turn the Grumman AA1-C from the base leg to final at the West Texas airport[22] T27 in El Paso, Texas. He is normally reserved, so his reaction shocks me. What am I doing wrong?

I had started the left turn from base to final a bit late, and coming out of the turn, we weren't lined up with the runway. I used left rudder to correct for this while keeping the airplane level with the ailerons. It seemed perfectly okay. So why would John shout, "Don't do that!"?

John explains when we are on the ground. The left rudder for lining us up with the runway plus the counteracting ailerons for keeping the plane level imply that we are flying with crossed controls. And that means that, if we get into a stall, we immediately enter into an unrecoverable spin and die. No wonder John was aghast.

What's to be learned?

The Lesson

No matter where you are flying, stay coordinated. For example, when turning left, use the left rudder *and* move the control stick to the left so that the plane banks into the turn. If you inadvertently stall the plane, you will not enter a spin, which would be unrecoverable if it happens close to the ground.

There are important exceptions. During a landing with crosswind, I prefer to compensate for drift by pointing the plane into the wind. Just before touchdown, rudder and opposite aileron then kick the plane around, as the saying goes, and thus line up the plane with the runway. That final step is a case of crossed controls. It isn't dangerous because you are close to the ground and a stall cannot produce major harm. I should mention that pilots endlessly debate whether this method or a cross-controlled side-slip during the entire final leg should be used.

A second exception is the forward slip to lose altitude. You make sure that you do not stall in that configuration by pushing the nose down.

The flight to El Paso was part of a trip to the Grand Canyon, Arizona. The purpose: Even after four years of flying, I always had queasy feelings about long cross-country flights. John suggested that we undertake a major trip over several days as a cure. Technically, he acted as instructor, but actually worked to build up my confidence.

That fateful trip to the Grand Canyon created many teaching moments. In particular, I learned how to handle adverse weather, including unexpected snow on the way from Prescott, Arizona to the Grand Canyon. The net effect: I built confidence in evaluating weather forecasts, planning routes, and dealing with unforeseen events.

That result prompts a suggestion: If you just have earned your pilot license, it is an excellent idea to take an extended cross-country trip lasting several days with an experienced pilot. They will act as mentor and instill in you the confidence that you can handle all steps on your own.

The next chapter also concerns landings, though in a different setting.

16

Landings

September 15, 2024

The winds at today's destination, the Flagstaff, Arizona airport KFLG, are 18 kts gusting to 24 kts. The ATIS also cautions about low-level wind shear, which means that the plane may suddenly pitch down while airspeed decreases, or it may pitch up while airspeed increases. The pitch-down case is nasty since the airspeed is reduced in two ways. First, by the pitch down and second since the pilot must raise the nose to counter the sudden descent. That two-fold reduction may well result in a stall. The pitch-up case is less troublesome: If due to the increased altitude and speed one cannot put down the plane on the remainder of the runway, one can always go around and try again.

How should I account for these conditions during the landing? As student pilot I learned the following customary rule.

- Divide the gust factor—the difference between gust speed and wind speed—by 2, and add that value to the normal landing speed.

In the Flagstaff case, the gust factor is $24 - 18 = 6$, and $6/2 = 3$ kts is added to the normal landing speed. Note that the rule doesn't account for low-level wind shear.

For three decades I relied on this rule even though the plane sometimes stalled just above the runway and settled with a bump. Chapter 3 describes an instance where Manfried and I landed at the Lancaster airport KWJF with very strong and gusty winds.

In 2010, the outstanding pilot and aviation writer Barry Schiff proposed the following alternate rule that avoids such botched landings.[23]

- When landing with severe wind conditions, maintain the groundspeed as indicated by GPS equal to the usual landing speed.

I use Schiff's rule for the landing at Flagstaff. The plane bounces around on final and for the first part of the runway. But as I react by raising or lowering the nose, the air speed remains comfortably above the stall speed, and the plane touches down smoothly.

Somebody may object to Schiff's rule with, "Don't the landings with that method require much more runway that is normally the case?" Not really. After all, when there is not wind at all, the landing speed is equal to ground speed.

Distractions in flight can be deadly. The next chapter covers a nasty case.

17
Distractions

September 4, 2016, 1 pm

Cruising westbound past the southern tip of the Guadalupe Mountains in West Texas, a surprising view of the Salt Flats next to the mountains unfolds. Normally, the Salt Flats are parched. But due to heavy rains in West Texas, they are wet and even contain some ponds.

Salt Flats with ponds west of the Guadalupe Mountains, Texas

Out comes the camera. I disengage the autopilot, dip the left wing for a good shot, use the telephoto feature of the camera, and wait for the right composition. I am totally absorbed.

Time flies when you are having fun, or here, when you are totally focused on taking a good picture. By chance I look up from the viewfinder. What the ...? The plane is spiraling downward at a steep angle. Ten more seconds of photography fun and I would have been in serious trouble, trying to recover from a deadly downward spiral.

The Lesson

If you are flying solo, as I did on that trip, you should not take photographs from the cockpit without engaging the autopilot. The reason is simple. To get a decent shot, you almost always have to dip or raise one wing. If an autopilot is not used, that sets you up for a deadly spiral since you are focusing on the photography. Instead, set up the autopilot for a constant turn, and then take photos as interesting scenes come into view. Hence I now have the following rule, which is yet another reason for having an autopilot.

- Taking photographs in the air must *always* be done either by the copilot or, if it is a solo flight, with engaged autopilot. No exceptions.

The next page shows a second unusual photo taken during that trip. Of course, this time with engaged autopilot. The extensive rain has produced rare vegetation in the desert connecting the Guadalupe Mountains and El Paso, and several hills display unusual contour lines.

You would never see this driving or flying commercially.

74 DISTRACTIONS

Hills with contour lines near El Paso, Texas

Landing at some airports with modest winds and clear sky can be hazardous. Read on to find out why.

18

Dangerous Airports

July 28, 2014, 12 pm

Immediately after touchdown on the runway of the airport KGIC in Grangeville, Idaho, the plane is rapidly shaken up and down as if struck by a giant sledgehammer. Manfried and I see the cause of the destructive movement once we have left the runway. Each individual concrete slab of the runway is significantly lifted up at one end. The result is a staircase structure where each raised slab end produces a terrific jolt. In hindsight, I would estimate that each jolt must be close to the g-force limit for the airplane. I am totally shocked by this horrific landing. As we shall see, it causes extensive damage.

Before going into details, I should mention that in 2024, 10 years later, the surface of the runway of airport KGIC is listed as asphalt in excellent condition. I don't know when the change from the disastrous concrete slabs to asphalt was made. No matter, landings can now be safely conducted.

The morning after that awful landing, Manfried and I fly from the airport KGIC to the Brigham City, Utah airport KBMC. For the takeoff we use a short part of the runway that doesn't have the staircase structure. The landing at Brigham City is uneventful. All seems to be well. That assessment changes the next morning.

July 30, 2014, 7 am

While preflighting the plane for the planned trip from Brigham City to Rawlins, Wyoming, I touch the muffler and realize that I can move it back and forth. Oh my! The welds of the two exhaust pipes holding the muffler have cracked!

After much discussion with the airport manager and a search for help far and wide, we locate a small machine shop willing to come to the airport with welding equipment and reweld the pipes.

The flight back home from Brigham City is uneventful. After each landing, I check the welds and confirm that they are still holding up. But after landing at the home airport T31 in McKinney, Texas north of Dallas, I once more check the welds: They have cracked again!

The Zenith factory in Mexico, Missouri, suggests that we bring the muffler and have them redesign it, including the attachment to the exhaust pipes. Some other modifications are still needed, but eventually the muffler works flawlessly.

On to the next problem caused by the horrible landing at the Grangeville airport.

June 12, 2019

During the annual inspection, a bottom flange of the motor mount turns out to be cracked, typical damage from high g-force jolts. The Zenith factory not only repairs the flange, but adds gussets to the engine mount for additional strength.

Two years later, yet another failure surfaces.

April 19, 2021

The annual inspection reveals that the bottom of the tailcone is cracked crosswise in two places. It's as if somebody had cut the sheet metal. It could only result from extraordinary g-forces pushing up the plane. I cover the cuts with overlapping sheet metal strips. The tail cone is now stronger than ever.

Analysis

At this point, you might be wondering if all of these problems were necessarily caused by a single horrible landing.

The counterargument: My landings are sometimes firm, but none of them remotely like the hammering of the plane with high g-forces experienced at Grangeville. I also never land on grass strips. That leaves possibly turbulence in flight. Chapter 5 describes the worst case, where Manfried and I encountered virga rain. In that situation the plane was tossed around, but never with the g-force of the hammering during the landing.

Conclusion

Before you land at any airport, always examine all key information: the availability of fuel, NOTAMs (NOtices To AirMen), the condition of runways, and so on. Apply the following ironclad rule.

- Never land on a runway that isn't in good or excellent condition unless there is no other choice.

I should add that this rule wouldn't have prevented the disaster at KGIC since the runway had been listed to be in good condition. Otherwise we wouldn't have landed there. That doesn't mean that the above rule is useless. Indeed, it appears that the classification of runways is now done more critically than in the past, and the stronger standard likely classifies any concrete runway with pronounced staircase structure of the slabs as being in fair if not poor condition.

Another takeaway: When one thinks about high g-forces damaging the plane, flights where sudden maneuvers or wind turbulence overstress wings or control surfaces come to mind. But high stress due to ground maneuvers can be just as damaging, for example, landings on a rough grass strip or, in our case, on a badly flawed runway. The latter situations may not matter much for planes with soft gear suspension or oversized tires. But when the suspension is fairly stiff and the tires are small, as in our plane, the jolts may exert big g-forces that cause major damage.

The list of disasters caused by the single horrible landing at the airport in Grangeville isn't complete, as we see in the next chapter. It starts Part III.

Part III

Emergencies

19

Exhaust System

June 28, 2022, 11 am

Ten minutes after takeoff from the airport KLXT of Lee Summit, Missouri, the engine emits a loud continuous blare. Even with my extensive experience with combustion engines, I have no idea what is wrong, and the first thought is, "The engine is breaking up." Now what?

Puzzling Symptoms

A strange situation has preceded the blaring sound. Shortly after takeoff, the coolant temperature climbs to 230 F. This is odd. The temperature usually hovers around 210 F and goes to 230 F only during a steep climb. I reduce power. The temperature goes down to 210 F within 30 seconds. The speedy temperature reduction is unexpected, too. It normally takes several minutes to cool down the engine by a reduction of power.

Adding power, the engine emits the loud blare again. I reduce power. The blaring is reduced. Adding power, the blaring sound becomes very strong. In fact, it's much worse than before. Clearly something under the cowl is … what? Breaking up, falling apart, blowing up?

Oil pressure and temperature are normal. The water temperature is in the green arc. But the engine sound is frightening. Responding intuitively, I reduce power to the point where the plane just stays at altitude, around 2500 ft AGL. Maybe the engine is about to seize up, maybe something has broken off, who knows?

Taking Action

The thought surfaces that I likely will be forced into an emergency landing. I look for a suitable meadow or road. Nothing appealing is in sight. A thought I never had before but have prepared for: It's time to call for help. I dial the emergency frequency 121.5 and broadcast, "Mayday, Mayday. This is Experimental N314LB with engine about to fail."

Within a few seconds, Air Traffic Control (ATC) responds and asks me to verify the call sign and provide the location. I have already prepared for this and provide the identifier, radial, and distance for the nearest VOR, even say that I have ADS-B Out, something ATC likely knows already.

ATC provides a transponder code, which I dial in. Then ATC points out options. There is an Interstate to the right, but with dense traffic. Farmed fields with much vegetation. ATC also mentions two airports, one 5 miles away, the other 10 miles. I am too focused on the engine to look up the airports, in fact do not believe they can be reached.

I emphasize to ATC that the engine may break up at any time. By now I am just 1500 ft AGL and concentrate on possible landing sites if the engine quits or worse. Given that mental preparation for an emergency landing, I just don't feel that I can spend time on airport selection. Then I have the correct idea: ATC has all the information. I just should ask for it.

I request details about the two airports. The closer one, KEZZ, is 5 miles north in Cameron and can be reached with a tailwind.

That makes it the obvious choice. I don't want to figure out the heading to the airport since I want to remain focused on emergency landing sites. Instead, I make another smart decision. I just ask for the direction to steer. ATC gives the information in a snap, and I swing around to that direction.

Upon another question, ATC provides the time required to reach the airport: four minutes. It's a long time when a complete engine failure seems imminent. To make it even worse, there are now trees below, and going for the airport might be the wrong decision. But I am committed now. The engine is still turning at the reduced rpm. That gentle treatment seems correct even though it extends the time to reach the airport. I pass over the trees and get to an expansive grass area surrounding the airport runway. Relief: I could land almost anywhere.

ATC gives the winds of the Kansas City airport even though the airport I am approaching has AWOS (Automated Weather Observing System), as I learn later. That system happens to be temporarily out of service. But there is the windsock. I tell ATC that the wind looks fine with a moderate crosswind component, something I can easily handle. I land, taxi off the runway, and tell ATC that I have done so. Then I realize that ATC cannot possibly receive that broadcast since I am too low for line-of-sight communication.

What do you know? There is a response. Incredulous, I ask how this is possible. As I learn later, ATC had asked another pilot to track my plane. He was nearby when I made the first Mayday call, and ATC asked him to keep watch over the plane. When direct contact with ATC was no longer possible, he began relaying my transmissions. Wow, what a system! I thank ATC profusely. What a terrific help when a complete engine failure loomed!

It seems that I handled everything right, doesn't it? Not quite. Let's see how the process looked from ATC's position, and at the same time figure out what I should have done differently.

Subsequent Contact with ATC

Three weeks after the fateful flight, Lisa of Kansas City ATC contacts me. She is the Air Traffic Controller Association president for Kansas City International Tower and Approach Control. She tells that several people had been involved when I called for help on 121.5. Up to that time, my assumption had been that just one controller had assisted. It actually had been a complex operation. She proposes a Zoom session with all the people who had been helping during that fateful flight.

Let's look at the flight from the viewpoint of ATC. The narration is based on a summary by Lisa. Comments in parentheses tell what happened in the cockpit.

The ATC Story

Mid-morning of that fateful day, Kansas City Center advises ATC that an aircraft has called on the Mayday frequency 121.5 in Kansas City airspace. The departure controller reaches out on that frequency and locates the aircraft in the Kansas City airspace south of the Cameron airport KEZZ. The controller requests that the pilot switch to 118.4 for better communication.

(I hear that request. By mistake, I just have pushed the button for switching frequencies, going back to the frequency for opening the flight plan with Flight Service after the departure from Lee's Summit. Confused, I think that I am talking with Flight Service, and decide to go back to 121.5 and continue there. This is a mistake, of course. I should have asked for clarification of the frequency information.)

The fact that the pilot continues on 121.5 posed a problem since it demands that the controller handle just that single aircraft. This requires a major reassignment of ATC personnel, including a controller who is on break. It involves initially two additional controllers, and later yet another one.

(All this is unbeknownst to me: I focus on the terrain below for possible emergency landing sites and communicate on 121.5.)

As luck would have it, a VFR aircraft has just departed from KGPH, one of the so-called satellite airports in the larger Kansas City area. The controller requests that the satellite controller responsible for that airport contact the pilot and request that the plane act as spotter aircraft. The pilot agrees to help, and the satellite controller sends him toward the aircraft in trouble. It's a terrific decision. In case of a crash, the spotter aircraft can immediately call in the precise location and thus direct rescue personnel.

(I am not aware of any of this.)

The controller acts cool and reassuring. He advises that there are two nearby airports, Cameron KEZZ and Midwest National KGPH.

(I completely discard that information, feeling that the engine will never last long enough to get to either airport.)

He also points out that Interstate I35 is just off to the right.

(Looking to the right, I feel great. This is the first landing possibility that seems doable without a crash.)

He repeats the advice about the two airports, seeing that the plane is just making circles in a search of an alternate landing site.

(I am caught up in the belief that the engine can quit at any moment, and keep looking for a landing site. A mistake, of course. The gentle prodding of the controller to think about the two airports finally gets me off the notion that I will never make it there. This is the crucial point in time when I ask for details about the two airports.)

He then supplies information about the two airports, including the distance, of course.

(It's at this point that I decide to offload all information gathering to the controller. It keeps me focused on alternate landing sites, without the need to fiddle with the radios and looking for relevant information.)

A second controller gathers the information the pilot has requested, and the controller passes it on. The information includes current distance, wind conditions, and direction to steer, and a bit later the time needed to get there.

(What seems to me an amazingly quick response by the controller actually is the result of two controllers working together. The wind conditions tells me that for any reasonable landing I will have to go around the airport and approach the runway from the north.)

The controller checks for competing traffic near the airport. Fortunately there is none, and hence no need for the pilot to switch to the airport frequency and announce the arrival and pattern legs.

(The thought doesn't even cross my mind that I should switch frequencies. Instead, I just look around carefully and check whether any other aircraft are in the area. There aren't any.)

While all this goes on, the spotter aircraft tracks the plane from above. When the plane is too low to connect with ATC via 121.5, the pilot of the spotter aircraft relays the transmissions to the controller. That includes telling ATC about the successful landing and taxiing to the FBO terminal.

When I learn about all this in the Zoom session with all personnel involved, I am just amazed: Four controllers and one pilot worked diligently to assist. The word "grateful" doesn't begin to express the thoughts I have every time I go over this process again.

Insight from the Zoom Session

When things really go south and you realize that help is needed, don't hesitate to launch a Mayday call on 121.5. There isn't just one person who might help you but an entire organization. They assist with things you aren't even considering, such as a spotter plane tracking your flight.

Do not believe that you will stay totally cool and rational. It may look like that to you, but you may overlook major items, both in your favor and against you. The controller will objectively look at your problem and may give you advice you haven't even thought of, let alone evaluated.

That was the case for me. I was so caught up with the notion that the engine could quit at any time that I ruled out trying for any airport. This was based on my understanding of the Rotax engine, the sounds it may make, the way it behaves when things are not normal. That understanding did not include the current case, as becomes evident in a moment.

Let's go on with the story. I have just landed at the airport KEZZ in Cameron, Missouri, taxied to the FBO, and shut down the engine.

What's Wrong With the Engine?

Eric of the FBO is most helpful. He allows me to move the plane into a hangar that has just been vacated. Nephew-in-law Ryan flies in from Ames, Iowa with the family's Cessna 150. He brings tools. Before he arrives, I have already looked over the engine, even have run it once more. Still the blaring sound, yet nothing seems amiss. I quickly shut down the engine.

With Ryan's tools, we pull the spark plugs and check for compression. We do not have a gauge, but sealing the opening for one plug at a time with a finger and turning the propeller by hand, we have the sensation that there is reasonable compression on all cylinders.

Ryan suggests that we run the engine for a few moments and check whether all exhaust pipes are hot. Lo and behold, the exhaust pipe of the #3 cylinder feels comparatively cool. Not only that, as Ryan touches the pipe, he also moves it! We look closely. The pipe has separated from the cylinder! Since the pipe is well hidden by the right-hand-side carburetor and the water radiator, I didn't see the problem when I first inspected the engine.

Broken exhaust pipe of cylinder #3

How is such a failure possible, given the careful manufacturing of that part by the Zenith factory? In my mind there is just one explanation: The incredible g-force jolts the plane encountered during the landing in Grangeville weakened the exhaust pipe and eventually caused it to break.

Analysis

The pieces of the puzzle come together. When the exhaust pipe is disconnected from a cylinder, that cylinder runs hotter due to a certain interaction between the fuel/air intake and the exhaust gas.[24] When we removed the spark plugs, we saw the effect of the increased temperature: The spark plug of the #3 cylinder had run hotter than usual, even at the low power setting.

The Rotax 912 engine originally didn't measure water temperature directly, but instead sensed the head temperature of one cylinder and declared it to be the water temperature. Our 30-year-old engine is of that vintage. The temperature probe happens to be installed on the #3 cylinder, exactly the one with the failed exhaust pipe. This explains the increased water temperature reading I had in level flight, and also that the temperature dropped almost immediately when I reduced power. Indeed, the water was still at 210 F, but the aluminum head of the #3 cylinder had gone up to 230 F. When I reduced power, the water quickly cooled the head back down to 210 F. The blaring sound resulted from a partial discharge of the hot exhaust gas past the broken pipe.

It almost certainly was a life-saving decision to immediately reduce power, since the escaping hot exhaust gas is like the flame of a propane torch. At any higher power setting, it likely would have damaged the carburetor directly above, and the resulting fire inferno would have been deadly. Babying the engine saved the day, as did the terrific ATC response to the Mayday call.

Repair

The Zenith factory in Mexico, Missouri helps. Roger says that they do not have the pipe in stock, but Travis produces a short portion connecting the pipe to the cylinder and ships it overnight. Ryan and I use it to repair the exhaust pipe, fly with the Cessna 150 to Cameron, install the part, test the engine, and all is well.

Repaired exhaust pipe

What's to Be Learned?

Prior to the Zoom session with the controllers, I had come up with the following three-part advice.

- When things seriously go south, call for help immediately on the emergency frequency 121.5.
- Rely on instinct when you react to the failure.
- Stay focused on the biggest threat, and ask ATC to figure out details of any plan and give you easy-to-follow step-by-step instructions.

The Zoom session with the ATC controllers led to additional conclusions.

- Do not hesitate to ask for clarification of ATC information, for example, about radio frequencies.
- Anticipate that your focus on one or two aspects will blind you to other problems and choices. Let ATC help you clarify those parts.
- Do not disregard ATC advice because you think it is irrelevant for the emergency. Instead, consider it carefully since you may be wrong in your analysis.
- Offload as many of the tasks to ATC. They are cool and focused, while you may overlook critical aspects.

A Well-deserved Award

In September 2023, the team of controllers who managed and co-ordinated the rescue effort received the 2023 Archie League Medal of Safety Awards: Central Region. It's wonderful that they received this recognition for their excellent work.

———

The exhaust pipe failure hasn't been the only near-disaster in my flying career, as seen next.

20

Burning Smell

May 18, 2004, 8 am

"Do you smell something burning?" I ask Manfried two minutes after an early-morning takeoff from the Merced, California airport KMCE. He responds, "Yes, I do."

Rationalization is my first reaction. "We must have gotten a few drops of oil on the muffler when we changed the oil. That's producing the smell," I tell Manfried. Then a scarier thought surfaces. Could this be a warning sign of a major problem that might end up with flames shooting out of the engine cowl? That arguments wins. I swing around, head back to the airport, fly the pattern, land, and taxi to the FBO.

The Problem

Manfried and I remove the engine cowl and carefully look over the engine. Oh boy, big problem: The safety wire of the plug at the bottom of the engine is broken and part of it dangles down. That plug holds the banjo fitting for the oil return line to the oil tank. Drops of oil on the plug indicate that it's leaking. In flight, those drops were blown onto the muffler and produced the burning smell. Using a wrench borrowed from the mechanic at the FBO, I discover that the plug is loose. If it had come off, the engine would have lost all oil in a matter of seconds and would have seized.

Thoughts tumble in my mind: Lucky that we returned to the airport; crazy that a safety wire can break; a huge mistake that I didn't see this during the oil change when I drained the oil tank and refilled it with new oil.

I ask the helpful mechanic whether he would be so kind to supply safety wire for the plug and the pliers to twist it. He comes with wire of the same gauge that had been used. "No, no," I object, "I would like to use a heavier gauge." Fortunately he has that, too. The rest is easy. I tighten the plug and secure it with the larger-gauge wire. The mechanic refuses payment, but I insist that I need to express my gratitude.[25]

Here's what I have learned.

Conclusions

- When symptoms during a flight indicate a potentially serious problem, skip over all rationalizing arguments claiming that it likely isn't bad. Instead, land at the nearest airport and investigate.
- If the problem is potentially life-threatening, forget about flight pattern rules when approaching the airport. Instead, merge directly onto final and land.
- Whenever you open up any part of the airplane at home or on the road, view this as an opportunity to look things over. Do this very carefully. It is helpful to say out loud what you are checking. For example, "Let's trace the fuel lines starting at the firewall."

In the situation at hand, I didn't follow the second rule. Indeed, I flew the complete standard approach for an airport: entered downwind, turned base, and then turned final. As I got on final, I realized this mistake and told myself, "What kind of fool am I?

Potentially, Manfried's and my life are at stake, and I am sticking to rules prolonging the flight!"

I also didn't follow the third rule. After the oil change involving the oil tank I should have gone over the engine and checked everything. If I had done so, I would have seen the damaged safety wire dangling from the plug at the bottom of the engine.

A third emergency happened early in my piloting career when I was still flying the Grumman AA1-C plane.

21

Almost Out of Fuel

Aug 21, 1989, 12 noon

Daughter Ingrid and I are flying with the Grumman AA1-C from Iowa back to Dallas. The airport KGLY in Clinton, Missouri, is planned as the first refueling stop. Headwind slows us down, but I convince myself that we can safely make it to the airport. This thought evaporates as we approach Clinton: Smack on top of the airport, a thunderstorm pours down buckets of rain, accompanied by lightning. We are approaching the endurance limit of the plane. What's to be done?

Choices

There is another airport nearby, but the AOPA airport book cautions that the runway is in bad condition. Other airports are some distance away and seem a poor choice compared with just waiting for the thunderstorm to move on. Fortunately, I had seen a recommendation about flying the Grumman when fuel runs low. It's time to use it.

Low-Fuel Management

When fuel runs low in a plane that uses one tank at a time—such as the Grumman—do *not* allow the fuel to become marginal in each tank. Indeed, if that happens, any selected tank may become unported, and the engine quits. This scenario is likely when you approach an airport and execute several turns. Here's a rule that avoids that calamity.

- When fuel begins to run low—in the Grumman AA1-C, when 30 minutes of endurance are left in each tank—run one tank at a time until it is completely dry.
- During that process, turn on the fuel boost pump and monitor the fuel pressure continuously. If you have a copilot or passenger, ask them to totally focus on the fuel pressure gauge and to warn when the fuel pressure begins to go down. At that point, switch to the next tank.
- Eventually all tanks except one are dry. At that time, that remaining tank still has a reasonable amount of fuel left that hopefully suffices to reach the destination airport.

In agreement with these rules, I switch to the right tank and step slightly on the left rudder to raise the right wing a bit. This assures that the tank will be completely drained. Then I ask Ingrid to totally focus on the fuel pressure gauge. I also slow down the plane to minimize fuel consumption, using just enough power to stay aloft, then fly back and forth sufficiently away from the thunderstorm. After 20 minutes, Ingrid announces the fuel pressure is falling. I switch to the left tank: We now have something like 30 minutes left. Gradually the thunderstorm moves away from the airport, and I land on a sopping wet runway.

As the lineman at the FOB fills the right tank, he says, "Wow, the tank must have been completely empty!" I respond, "That was the plan." The left tank has some fuel left, maybe for another 20 minutes.

All this happened many years prior to ADS-B. Today, I would have become aware of the thunderstorm while en route due to Sully's Rule and would have worked out an alternative.

The trip influenced the design of the fuel system for the Zenith 601HDS. When Mel and I built the plane, we added an electric fuel pump to each wing tank. These pumps feed a center tank, which in turn supplies the fuel to the Rotax engine. The transfer from the wing tanks to the center tank is part of Sully's Rule, and hence is carried out every 30 minutes.

On a long-distance flight, the two wing tanks become empty after 3 1/2 to 4 hours. Typically I arrive at the destination before that happens, but in rare cases I must go on. The center tank then has fuel left for two hours. I allow myself to use at most half of that remaining fuel before landing, thus always arrive with at least one hour in reserve. That's a comfortable margin even in demanding situations, for example, when flying across mountains or desolate areas.

In my opinion, it is essential that each pilot understands how all that technology under the engine cowl and in the cockpit works. An unreasonable demand? Let's see.

22

Excessive Oil Pressure

June 16, 2009, 6 pm

After refueling at the Claremore airport KGCM near Tulsa, Oklahoma, I take off for Dallas. Two minutes into the flight, the oil pressure gauge shows high and fluctuating values. I hesitate for a moment, but then caution wins. I swing around and land again at the airport. Contrary to published information, the airport is unattended. Literally nobody is there to help. Now what?

Analysis

I remove the top part of the cowl and look over the engine. All appears to be as expected: There is enough oil in the oil tank, there is no leak, the cable connecting the oil pressure sensor with the pressure gauge is firmly attached at the sensor.

Another runup of the engine. It sounds and acts as expected, but there are the strange spiking values of the oil pressure gauge. My conclusion: This is either a failure of the oil pressure sensor or of the gauge. It is supported by the fact that a ball and spring in the oil pump control maximum oil pressure. That simple arrangement is about as reliable as could possibly be.

The takeoff and flight back to Dallas are uneventful, except that the time spent for the investigation at the KGCM airport delays the flight enough that I must stay overnight in Sherman, Oklahoma just 30 minutes short of my home airport T31 in McKinney, Texas.

The next day I check in with Jack, the resident guru of T31. He says that my reaction was correct. Since the pressure indication was high while everything else looked normal, the pressure sensor most likely had failed. This turned out to be the case.

Conclusion

I felt helpless at Claremore Regional after I had landed and nobody was there to consult with. Hence, I did as well as I could, checking the engine and its operation carefully, and concluding that either the oil pressure sensor or the gauge were at fault.

When I was back home, the events of that day prompted me to carefully look over all sensors and gauges and understand how and why each of them could fail. Here is my advice for that process and use of the insight.

- Collect information about each sensor and instrument in your plane, in particular, how they work and how they can fail. Ask experts when you are not sure about some of the possible situations.
- When indicated values are not as expected, land and investigate carefully. Use your preparation to arrive at a diagnosis, then handle accordingly. If in doubt about your diagnosis, pull in experts to help you.

As part of the first step, I examined all possible situations involving wrong oil pressure indications for the Rotax engine. The analysis brought to light that some failures are significant while others have no impact on safe flight. A few years later I published the results in a detailed blog post.[26] Another post covers cases where coolant mysteriously disappears from the Rotax engine.[27]

The next chapter concerns an event you hope will never happen.

23
Pilot Incapacitation

It's fortunate that I am unable to start this chapter with an instance where I became incapacitated while flying high up. But it behooves every pilot to plan for this event by preparing a copilot or passenger to take over.

A small checklist on board helps with this process. It should be specific for the aircraft, of course. In my case, it is the following list.

Checklist for Pilot Incapacitation

1. Stabilize airplane to straight and level.

2. Check fuel gauge on left hand side and estimate endurance, using 4 gal/hr as fuel burn rate.

3. Find nearest VOR by turning right small knob of Bendix King KLX 135A radio until page of the nearest VOR appears.

4. Change transponder code to 7700 and press "ident."

5. Set standby radio frequency to 121.5 using left large and small knobs of Bendix King KLX 135A radio. Press small left button to activate the frequency.

6. Press right hand side white button near throttle and broadcast, "Mayday, Mayday. This is Experimental November 3-1-4 Lima Bravo (N314LB). We have incapacitated pilot. We are on the (number) radial of the (identifier) VOR at (distance) miles. Altitude (read altimeter), speed (read air speed indicator) kts, endurance (give hrs)."

7. Carry out instructions while keeping altitude, airspeed, and attitude under control.

8. For landing, stabilize plane at 70 kts landing speed and 200-300 ft/min rate of descent. Reduce engine power to idle as threshold is crossed.

9. When plane has stopped, turn off all switches, turn off ignition, and leave cabin.

10. Help the pilot.

Instructing the Copilot/Passenger

Have the person take over the controls and carry out the above steps, of course without actually broadcasting information. You act as the controller who responds to the emergency call. To simulate the landing, you select an altitude and have the copilot/passenger act as if the runway was at that level.

And now you can relax. If something bad happens to you, the copilot/passenger can take over and safely land the plane at an airport. It likely won't be a pretty landing. But that's not the objective. The goal is survival.

The next part covers good practices for your plane and also yourself, that is, your body and soul.

Part IV

Good Practices

24
Maintenance

You or a certified mechanic must inspect your plane once a year using a rigorous process. If a mechanic is involved, it is important that you participate as much as possible. That way you learn many things you wouldn't know otherwise. But the yearly inspection does not suffice. If you haven't flown the plane for a number of weeks, I recommend the following check.

- Remove the engine cowl and inspect everything: the fuel delivery system, the coolant system, the electrical system. Pull the propeller through a few times to sense the compression strokes of the cylinders. Look at the exhaust system for any damage, include the heat muff mounted around the muffler.
- Look at the landing gear, check tire pressure, and inspect the brake system.
- Examine the fuselage and wings and move the control surfaces.

In short, do a reasonable inspection that all is well. Somebody may argue, "That's not really necessary. After all, there is the yearly inspection, and during the most recent flight everything worked fine."

Remember Murphy's Law "If it can go wrong, it will"? For aviation I prefer the stronger version "Everything can go wrong." The proposed inspection aims to outsmart that rule.

The next chapter talks about body and soul. If you already know everything about that topic, just skip the chapter.

25
Your Body and Soul

If you are just interested in flying airplanes for a few years, you can safely skip this chapter. But if you want to fly for many years, even multiple decades, you need to maintain your mental and physical fitness consistently, and may want to read on.

A multitude of books tell you how to achieve that goal. Pick whatever book works for you and implement its program. The key for success is sticking with the process. Of course, at some time you might change to another program that's even better. But then you carry out the new regimen with persistent effort, too.

As for me, people have asked time and again how I manage to fly an airplane at age 82, when most pilots have given up, and how I can accomplish physical tasks such as hiking up high mountains or into deep canyons. After several years of such inquiries I decided to write the book *Subconscious Blunders: A 21st Century Epidemic*[28] to share what has worked for me.

The title of the book is motivated by the fact that many ills of modern life stem from errors of the subconscious machinery operating within each of us. Indeed, multiple industries are exploiting weaknesses of that machinery to attain huge profits. You can escape their seductive manipulation through deliberate and persistent counteraction. Doing so, you achieve a healthy body and soul.

The chapters so far have shown that lots of things need to be considered for safe long-distance flight in a small plane. At times, you might have thought, "Why should I bother to learn so many rules? Why not drive a car or fly with a commercial airline instead?"

The chapters of the next part give an indirect answer to that question. They describe some wonderful trips with amazing views. Neither driving nor flying commercially can deliver this experience.

Part V

Joy of Flying

26
Mountains

Flying across mountains in a small plane requires care, as discussed in Chapters 3, 6, 7, and 8. But the rewards are extraordinary.

One of my favorite mountain flights approaches Grand Teton National Park from the northeast, starting from Riverton, Wyoming early in the morning. Flying up the winding road leading to the Togwotee Pass, the entire Teton mountain range suddenly comes into view.

Sometimes the winds are easterly, and I am on the upwind side of the Teton Mountains and can fly up close and peer into glacier-created valleys.

112 MOUNTAINS

Approaching Grand Teton National Park from the northeast via the Togwotee Pass

Glacier valley high up in the Teton Mountains

The Teton Mountains are just as impressive when seen from the west, here after an early-morning takeoff from Rexburg, Idaho.

Teton Mountains seen from the northwest

Planning the above flights was easy. But a crossing of the Rocky Mountains from the Montrose, Colorado airport KMTJ to the Denver area was more complex; see the green route below.[29] It largely

Route from the Montrose, Colorado airport KMTJ to the Denver area

is the reverse of a flight described in the terrific sci-fi book *The Dog Stars* by Peter Heller.[30] Montrose doesn't occur in the book. I selected the route to start there so the subsequent description in one of my blogs[31] wouldn't give away key parts of the book.

The views of the three hour flight are spectacular. A few minutes after the early-morning takeoff from Montrose, the Gunnison River appears on the right-hand side. It glistens like molten silver.

Gunnison River northeast of Montrose, Colorado

Mt. Sopris comes up next.

Mt. Sopris

Near the end of the flight, I climb to 14,000 MSL to be 2,000 ft above the Rollins Pass, then glide down into the Denver area.

At 14,000 ft MSL, 2000 ft above the Rollins Pass

You may wonder: Don't I require supplemental oxygen at 14,000 ft altitude? No, I do not. Almost every year I hike for hours in the

Colorado mountains up to 13,000 ft MSL without ill effects. Given that experience, I surely can operate the controls of the plane safely at 14,000 ft MSL up to one half hour, the mandated time limit for flight above 12,500 ft MSL without supplemental oxygen.

Caution

Some pilots require supplemental oxygen at high altitudes. If you are contemplating such a flight and are unsure what you can and cannot do, find out in a trial flight where you measure oxygen saturation content of the red blood cells with an oximeter. At the same time, check for symptoms of hypoxia such as rapid heart rate and shortness of breath.

We purposely omit here critical oximeter values since the effectiveness with which the body processes the oxygen of the red blood cells plays a central role, yet is not part of the oximeter measurement. The effectiveness is influenced, for example, by the way you breathe, that is, mouth breathing versus nose breathing,[32] and by the efficiency with which the body utilizes the oxygen once it has become part of the body cells.

Tracing rivers can be another terrific experience, as the next chapter shows.

27
Rivers

During the 1980s, Arie and I traced the entire route of the Lewis and Clark expedition[33] of 1804-1806 with the Grumman AA1-C: up the Missouri River, across the Bitterroot Mountains, and down the Columbia River to the Pacific coast. We took photos with an old-fashioned camera. Unfortunately, those pictures have faded by now. Much later, digital cameras became available and created permanent visual records of trips. Here are stunning vistas of two trips I took in 2016 and 2017 that trace the Dolores River and the Snake River from their headwaters to the confluence with a larger river.

The 241 mile-long Dolores River starts in the southwestern part of Colorado and flows northbound. After crossing into Utah, it joins the Colorado River. Access by car involves dirt roads, which I always avoid. Hence I had never seen the Dolores River during car trips going west. It was only by chance that I became aware of its beauty on a flight north from Albuquerque, New Mexico and decided to take a look. Following the meandering river northbound, I was amazed by the ever-changing rugged landscape.

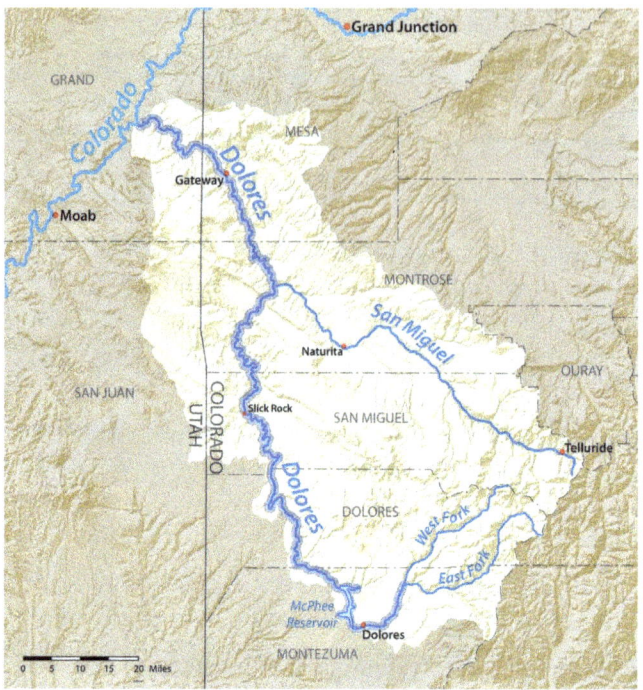

Dolores River[34]

Early section of Dolores River

Later section of Dolores River

The flight was a perfect argument for the claim that driving a car produces a 2-dimensional view of the world, while flying supplies 3-dimensional images. You might argue, "But photos of the flight *are* 2-dimensional." That's correct, of course. But the memory of the flight triggered later by the photos *is* 3-dimensional.

The Snake River is a wonder of nature. At 1,078 miles, it is one of the longest rivers of the US. It starts in western Wyoming. Southbound, the river flows along the eastern part of the Teton mountain range, swings around the southern edge of that range, and then proceeds northwest into Idaho. From there, it flows westerly in a long arc through that state. Finally, the river turns north, forming the border of Idaho and Oregon, and joins the Columbia River at Pasco, Washington.

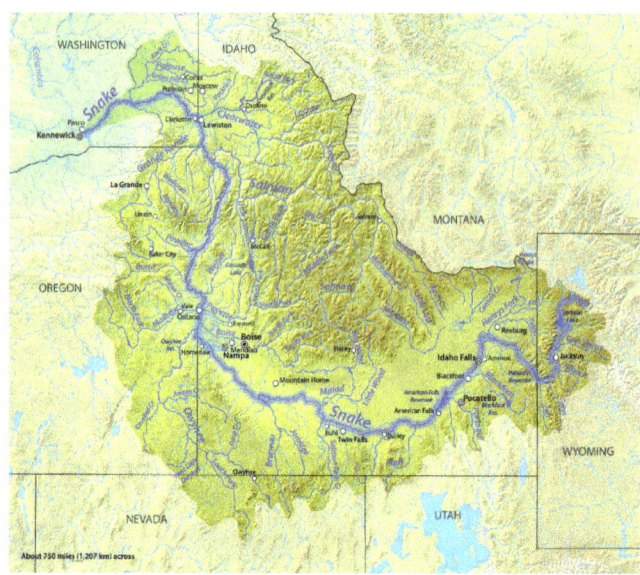
Snake river[35]

In Twin Falls, Idaho, the river creates stunning Shoshone Falls.

Shoshone Falls in Twin Falls, Idaho

Views of the Grand Canyon of the Snake River supply another argument that flying is 3-dimensional and driving 2-dimensional. Here are two photos. The first one shows the scale of the mountains surrounding the Grand Canyon of the Snake River; the river is nestled into the lower right corner. The second photo provides a close-up view.

Grand Canyon of the Snake River ...

The low-level photography required care. I used the autopilot to fly circles and as a precaution turned on the Terrain option of the Garmin Pilot to color portions of the map red (= don't go there), yellow (= caution is advised), green (= altitude is safe), and no added color (= no problem whatsoever).

The flights along the Snake River filled three days. You may wonder why it took that long. The idea was to slow down the plane as much as possible to extend the time gliding along the river. Small detours explored nearby terrain. These were ambling flights, so to speak.

Intrigued by these examples? Tempted to take such flights? Deciding to do so? Great!

You have reached the end of the book. Well, almost. There is still an epilogue that connects the book with brain science. Really? Yes, indeed.

Part VI

Epilogue

Mankind dreamed of flying for centuries. About 150 years ago, a determined effort started, first to understand how one can fly by starting from a high point, second to do so from level ground with an engine, and third to fly safely. The third part took many decades. It's only recently, with the advent of GPS, ADS-B, and elaborate weather analysis, that every pilot has extensive in-flight information available: about the route, obstacles, airports, weather, other nearby planes, and more. It is a wealth of information. At the same time, engines and electronics have become much more reliable.

However, these astonishing improvements do not affect some fundamental aspects of flying: The weather does whatever it wants; mountains create hazards; equipment fails. The result: There are still many aviation accidents involving small planes.

What's to be done? Both the AOPA and the EAA make a determined effort in their publications and safety programs to help pilots fly safely. Having flown small planes for 45 years, I wondered: Could I somehow assist in that effort? It isn't an idle question. Many aviation books already advise how to fly safely in a small plane, including two handbooks published by the FAA (Federal Administration Administration).[36] On the other hand, there still are all those accidents!

The thought surfaced that some results of modern neuroscience, popularly known as brain science, might motivate a format that would make the intended book more effective than a straightforward technical discussion. Here are the arguments.

All processing of the nervous system begins subconsciously. As part of the output, feelings and unbidden thoughts are sent to the conscious portion of the system, which considers that information and makes decisions.

There is a deceptive aspect. On the surface, it seems that the conscious part constitutes the essence of that activity. Hence one may argue that pilots can be convinced to fly safely by supplying material that carefully explains the conscious decision processes. But that conclusion is wrong. The bulk of the activity actually takes place at the subconscious level.

But there is hope. It turns out that properly selected conscious thinking *can change* subconscious processes. The idea, then, is to create material that in the reader triggers conscious thinking that changes subconscious processes. What must such material look like?

The theory behind CBT (Cognitive Behavioral (Therapy)[37] provides the answer. CBT is applied in psychotherapy with great success to effect subconscious changes with conscious thoughts. A key aspect is the following. The conscious thoughts must trigger strong emotions if they are to cause subconscious changes. That fact motivates how a book of safe flying should be written: It should provoke strong emotions in the reader. This book is based on that insight. Almost every chapter begins with a shocking paragraph that triggers emotions in the reader. Then comes the complete story, followed by recommendations how the particular problem or disaster can be avoided.

Two results of brain science have confirmed that CBT is effective when strong emotions accompany thoughts.

- Emotional words prompt release of dopamine, serotonin, and norepinephrine ([Batten et al., 2025]).
- Dopamine helps consolidate memories ([Wise, 2004]).

In the earlier terminology: Emotions produce dopamine, which in turn helps conscious thoughts change subconscious processes.

When the idea of this book first surfaced, I wondered how I could get many different stories that would fit into the intended chapter format. Then it struck me: During more than four decades of flying, I had experienced lots of difficult situations, indeed enough of them to assemble many chapters. Then came a second thought: How is it possible I got involved in so many difficulties, yet never had an accident? The motto cited at the beginning of the book provides the answer.

> *There are old pilots and there are bold pilots,*
> *but there are no old, bold pilots.*

Never was I bold. Often shocked, as you have seen, but never daring. Those shocks resulted in better subconscious processes, which in turn led to safer piloting decisions.

If you are a pilot, my hope then is that by reading this book you also experienced shocks and thought in detail about the cases. CBT says that these steps modified your subconscious processes and will lead to safer piloting decisions. A daring argument, isn't it?

Appendix A: VFR Flight

This book is about flying long cross-country trips under VFR (Visual Flight Rules) in a small plane. Roughly speaking, such flight cannot enter clouds. The book completely ignores flight under IFR (Instrument Flight Rules). Indeed, the Zenith 601HDS I have owned for the past 30 years cannot be certified for IFR flight, so it has never been an option.

If that sounds like a regret, be assured that it isn't. Flying for me means traveling, sightseeing, photographing, detouring on a whim, in short, enjoying the flight. IFR flight is very structured and hence of little interest to me.

In my opinion, the following equipment beyond basic instrumentation and radios is essential for a small plane flown VFR on long cross-country trips.

- ADS-B (Automatic Dependent Surveillance-Broadcast) In and Out. The system supplies essential weather information such as radar images and conditions at airports, and also displays nearby traffic. It is mandatory for many operations anyway.
- A flight management system such as Garmin Pilot or ForeFlight. The book always refers to Garmin Pilot since it has been my choice.
- A GPS-based backup system that suffices in case the flight management system fails.
- An autopilot providing at least two-axis control keeping the

airplane level and on course. Desired but not mandatory are altitude control or fancy features producing fully automated flight patterns.

- Paper charts, to be used if all electronic equipment goes south.

You may balk at the autopilot requirement, but for reasons laid out in several chapters we consider it very important for safe flight.

You may consider the paper charts requirement superfluous. Call me old-fashioned, but I cannot agree that everything depends on electronics. In fact, for each cross country flight I draw the route on paper charts, move along the route to identify possible difficulties, and circle items that require particular attention, such as 2,000 ft tall monster towers. That process not only forces me to examine details of the trip, but also stores important data in my brain that I recall during the flight.

Glossary of Terms

100LL	aviation gasoline containing 2 grams lead per gallon
ADS-B	Automatic Dependent Surveillance-Broadcast. Produces position and speed of the aircraft and nearby ones in the cockpit. Also supplies critical information about weather, airports, wind speeds, and more.
AGL	Above Ground Level
AIRMET	AIRman's METeorological information
AOPA	Aircraft Owners and Pilots Association
A&P	Airframe and Powerplant
ASOS	Automated Surface Observing System
ATC	AIR Traffic Control
ATIS	Automatic Terminal Information Service
AWOS	Automatic Weather Observation System
CA	Commercial Aviation: commercial air transport, in particular scheduled airline services, or aerial work
DAR	Designated Airworthiness Representative
DG	Directional Gyro
ETE	Expected Time En route
EAA	Experimental Aircraft Association
FAA	Federal Aviation Administration
FBO	Fixed Base Operator. Provides aviation services to the GA (General Aviation) community at an airport.

fpm	climb or descent rate, in feet per minute
FSS	Flight Service Station. Provides information and services for pilots before, during, and after flights.
GA	General Aviation: civilian, non-commercial flight
GPS	Global Positioning System
IFR	Instrument Flight Rules. They define how flight using instruments must be conducted.
kts	speed measured in nautical miles per hour = 1.15 statute miles per hour
MSL	Mean Sea Level
NOAA	National Oceanic and Atmospheric Administration
nm	nautical miles
NOTAM	NOtices To AirMen
SIGMET	SIGnificant METeorological information
STC	Supplemental Type Certificate. The document approves a specified modification of the aircraft.
TAF	Terminal Area Forecast
TBO	Time Between Overhauls
VFR	Visual Flight Rules. They define how flight using visual references must be conducted.

Notes

Wikipedia entries often supply an extensive list of references for additional explanations. Pointing to the Wikipedia entry avoids listing those references.
Better yet, as insight into a topic grows, the Wikipedia changes as well. Hence the reader always obtains the latest information about the topic.
All links were verified in October, 2024.

1. The Weiss airport was shut down on May 1, 1994. See Wikipedia "Weiss Airport."

2. VFR flight must avoid clouds. For details of the rules, see Wikipedia "Visual flight rules." A summary is included in Chapter 9. Appendix A specifies equipment of the plane that in my opinion is essential for VFR cross-country trips.

3. Source: Michael J. Hull, the current owner of the Grumman AA1-C N9782U, kindly supplied both photos.

4. Source: Mel Asberry kindly supplied the photo.

5. Source: Ingrid Truemper kindly supplied the photo.

6. Source: Ryan White kindly supplied the photo of Zenith 601HDS in the air.

7. For details about the maintenance of the Rotax 912 engine that produced the amazing longevity, see the post https://pointsforpilots.blogspot.com/2019/10/rotax-912-engine-2000-hours-in-25-years.html.

8. John Barrer kindly supplied the photo.

9. Unfortunately, I could not obtain a photo of Jack Wybenga.

10. See Wikipedia "Sully Sullenberger."

11. Source: "US Airways Flight 1549 as it floats on the Hudson River" by Greg L, available under Creative Commons BY 2.0. https://upload.wikimedia.org/wikipedia/commons/8/8f/US_Airways_Flight_1549_%28N106US%29_after_crashing_into_the_Hudson_River_%28crop_2%29.jpg.

12. See Wikipedia "Sully Sullenberger."

13. See Wikipedia "Sully Sullenberger."

14. Source: "C. B. 'Sully' Sullenberger" photo by United States Department of State, Public Domain. https://commons.wikimedia.org/w/index.php?curid=118219289.

15. VFR flight must avoid clouds. For details of the rules, see Wikipedia "Visual flight rules." A summary of the main conditions is included in Chapter 9. Appendix A specifies equipment that in my opinion is essential for VFR cross-country trips.

16. The post "Mountain Flying" available at https://pointsforpilots.blogspot.com/2006/01/mountain-flying.html summarizes the key concepts and considerations of flight across mountains.

17. The magenta area restricts VFR clear-of-clouds flights to at most 700 ft AGL so that IFR traffic coming into the airport has time to see any VFR aircraft below. There is a more severe version where the standard cloud condition is enforced from the ground up.

18. Source: "Typical thunderstorm over a field"' photo by Bidgee - Own work, CC BY 3.0, https://commons.wikimedia.org/w/index.php?curid=3799323

19. You could also use an appropriate chart or computer program to obtain the density altitude. The simple rule proposed here has the advantage that you need no tools.

20. The installation of the oil cooling fan is described in https://pointsforpilots.blogspot.com/2011/08/cool-oil-is-cool.html.

21. The increase is partially due to the insight that the Rotax 912 engine of my Zenith 601HDS measures cylinder head temperature and not water temperature, and thus overestimates water temperature.

22. The privately owned airport was created in 1977. In 2008, it was

renamed to Horizon Airport. It ceased operation in 2014. For details, see Abandoned & Little-Known Airfields: Texas: El Paso area https://www.members.tripod.com/airfields_freeman/TX/Airfields_TX_ElPaso.htm.

23. [Schiff, 2010].

24. For details about the interaction of fuel/air intake and exhaust gas, see the section "Valve overlap" of Wikipedia "Valve timing."

25. The broken safety wire had 0.032 in. diameter and the thicker wire 0.041 in. In later years, we never had another problem with the thicker wire. Amazing what a difference a thickness increase of 0.009 in. can make.

26. The post "Wrong Oil Pressure of the Rotax 912/914 Engines June 06, 2016" explains in detail the failure possibilities and appropriate actions. See https://pointsforpilots.blogspot.com/2016/06/wrong-oil-pressure-of-rotax-912914.html.

27. The post "Vanishing Coolant of Rotax 912 Engine" explains how coolant of the Rotax engine may disappear without an obvious leak. See https://pointsforpilots.blogspot.com/2017/07/vanishing-coolant-of-rotax-912-engine.html.

28. [Truemper, 2023].

29. The flight starts at 5,800 ft MSL in Montrose and rises to 14,000 ft MSL near the Denver area. During the gradual ascent, the route bypasses taller mountains while staying at least 2,000 ft above the terrain. In addition, the segment where the altitude goes beyond 12,500 ft MSL doesn't require more than 30 minutes in agreement with the rules for flight without supplemental oxygen.
The Garmin Pilot displayed the route, but I also drew it on the Denver sectional, with the planned altitudes of the flight added to the waypoints.
Finally, I scheduled the flight on a day with westerly winds blowing across the Rocky Mountains. This assured that the route was never on the turbulent downwind side of a looming mountain.

30. See [Heller, 2012].

31. For details of the trip from Montrose to the Denver area, including a comment by Peter Heller, see https://passionforflight.blogspot.com/2016/08/tracing-flight-of-beast-in-dog-stars.html.

32. The book *Breath* [Nestor, 2020] has details about the effectiveness of nose versus mouth breathing.

33. See Wikipedia "Lewis and Clark Expedition."

34. Dolores River map by Shannon1. Licensed under CC 2.5 Generic.

35. Snake River map by USGS and modified by Shannon1 - Terrain data from DEMIS Mapserver, CC BY-SA 3.0

36. The FAA handbooks are *Airplane Flying Handbook FAA-H-8083-3C* and *Aviation Weather Handbook FAA-H-8083-28*; see [Federal Aviation Administration, 2021] and [Federal Aviation Administration, 2022].

37. For details about CBT and its effectiveness, see Wikipedia "Cognitive behavioral therapy."

Bibliography

[Batten et al., 2025] Batten, S. R., Hartle, A. E., Barbosa, L. S., Chiu, P., Montague, P. R., and Howe, W. M. (2025). Emotional words evoke region- and valence-specific patterns of concurrent neuromodulator release in human thalamus and cortex. *Cell Reports* 44, issue 1115162, January 28, 2025.

[Federal Aviation Administration, 2021] Federal Aviation Administration (2021). *Airplane Flying Handbook FAA-H-8083-3C: Pilot Flight Training Study Guide*. Federal Aviation Administration.

[Federal Aviation Administration, 2022] Federal Aviation Administration (2022). *Aviation Weather Handbook FAA-H-8083-28*. U.S. Department of Transportation and Federal Aviation Administration.

[Heller, 2012] Heller, P. (2012). *The Dog Stars*. Alfred A. Knopf.

[Nestor, 2020] Nestor, J. (2020). *Breath: The New Science of a Lost Art*. Riverhead Books.

[Schiff, 2010] Schiff, B. (2010). Maintain Thy Groundspeed. *Pilot* magazine of AOPA, December 5, 2010.

[Truemper, 2023] Truemper, K. (2023). *Subconscious Blunders: A 21st Century Epidemic*. Leibniz Company.

[Wise, 2004] Wise, R. A. (2004). Dopamine, learning and motivation. *Nature Reviews Neuroscience* 5, June 2004, pp. 483-494.

Acknowledgements

I. Truemper and U. Truemper were patient editors. The University of Texas at Dallas—our home institution—made essential resources available.

We thank them for their help.

K. T.

Index

ADS-B, 20, 23, 42, 125, 128
AIRMET, 18
airplane
    Grumman AA1-C, 2, 5, 56, 67, 95, 117
    Zenith 601HDS, 6, 59, 60, 97, 128
airport
    3F6 Richards, Texas, 46
    3WE Fenton, Missouri, 1
    C17 Marion, Iowa, 1
    KAEG Albuquerque, New Mexico, 46
    KAKO Akron, Colorado, 53
    KBMC Brigham City, Utah, 75
    KBRL Burlington, Iowa, 2
    KCDS Childress, Texas, 46, 52
    KCOD Cody, Wyoming, 35
    KCYS Cheyenne, Wyoming, 27, 35
    KEAN Wheatland, Wyoming, 36
    KEZZ Cameron, Missouri, 82
    KFLG Flagstaff, Arizona, 70
    KGCM Tulsa, Oklahoma area, 98
    KGIC Grangeville, Idaho, 75
    KGLD Goodland, Kansas, 53
    KGLY Clinton, Missouri, 95
    KGXY Greeley, Colorado, 31
    KHHF Canadian, Texas, 52
    KIFP Bullhead City, Arizona, 60
    KITR Burlington, Colorado, 53
    KLXT Lee Summit, Missouri, 81
    KMCE Merced, California, 92
    KMHV Mojave, California, 15
    KMLB Melbourne, Florida, 49
    KMTJ Montrose, Colorado, 113
    KPVW Plainview, Texas, 46
    KRIW Riverton, Wyoming, 29, 31
    KSAF Santa Fe, New Mexico, 56
    KSET St. Charles, Missouri, 2
    KTTC Tucumcari, New Mexico, 56
    KWJF Lancaster, California, 16
    T27 El Paso, Texas area, 67
    T31 McKinney, Texas, 76, 99
Automatic Dependent Surveillance-Broadcast, *see* ADS-B
autopilot (importance), 42, 54, 73, 128

Forecast Evaluation, 47, 50, 54

Grumman AA1-C, 2, 5, 56, 67, 95, 117

hot weather, 57, 60

KMHV Mojave, California, 15

Leibniz, Gottfried Wilhelm, 6

rain, 1, 42, 44, 53

SIGMET, 53
Sully's Rule, 23, 29, 32, 50, 51, 54, 97

thunderstorm, 3, 42, 49, 53, 95

virga rain, 26

weather
  ADS-B, 20, 23, 42, 125, 128
  AIRMET, 18
  Forecast Evaluation, 47, 50, 54
  hot, 57, 60
  rain, 1, 42, 44, 53
  SIGMET, 53

Sully's Rule, 23, 29, 32, 50, 51, 54, 97
thunderstorm, 3, 42, 49, 53, 95
virga rain, 26

Zenith
  601HDS, 6, 59, 60, 97, 128
  factory, 76, 89

www.ingramcontent.com/pod-product-compliance
Lightning Source LLC
LaVergne TN
LVHW021559070426
835507LV00014B/1866